Error Patterns in Computation

Robert B. Ashlock

Covenant College
Lookout Mountain, Georgia

Prentice Hall
Englewood Cliffs, NJ 07632

Library of Congress Cataloging-in-Publication Data
Ashlock, Robert B., 1930–
    Error patterns in computation / Robert B. Ashlock. — 6th ed.
        p.   cm.
    Includes bibliographical references.
    ISBN 0-02-304212-5
    1. Arithmetic—Study and teaching (Elementary)   I. Title.
QA135.5.A782   1994
372.7′2044—dc20   93-18440
                                                        CIP

Cover art: The Image Bank/Pierre-Yves Goavec
Editor: Linda James Scharp
Production Editor: Sheryl Glicker Langner
Art Coordinator: Lorraine Woost
Text Designer: Jill E. Bonar
Cover Designer: Cathleen Norz
Production Buyer: Patricia A. Tonneman

This book was set in Garamond by American-Stratford Graphic Services, Inc.

 © 1994 by Prentice-Hall, Inc.
A Simon & Schuster Company
Englewood Cliffs, New Jersey 07632

Earlier editions copyright © 1990, 1986, 1982, 1976, 1972 by Merrill
Publishing Company.

Printed in the United States of America

10 9 8 7 6 5 4 3 2

ISBN   0-02-304212-5

Prentice-Hall International (UK) Limited, *London*
Prentice-Hall of Australia Pty. Limited, *Sydney*
Prentice-Hall Canada Inc., *Toronto*
Prentice-Hall Hispanoamericana, S.A., *Mexico*
Prentice-Hall of India Private Limited, *New Delhi*
Prentice-Hall of Japan, Inc., *Tokyo*
Simon & Schuster Asia Pte. Ltd., *Singapore*
Editora Prentice-Hall do Brasil, Ltda., *Rio de Janeiro*

Preface

INTRODUCTION

This book was written for those who want to help every student learn to compute successfully. It is designed to help readers become more sensitive to what a student does, to learn from thoughtful analysis of a student's papers and from careful listening to what the student says.

Specifically, this book is about detecting systematic errors students make when doing computation; reasons students may have learned the erroneous procedures are also examined. Strategies for helping those students are included. Of course, many of the instructional strategies are useful when teaching any student—whether the student has experienced difficulty under previous instruction or not.

An attempt has been made to deal with real problems in a realistic manner. The patterns of error in computation that are included in this book are not figments of the author's imagination; they are observed patterns used by real boys and girls—usually children in regular classrooms. The problems in diagnosis posed for the reader are typical of difficulties encountered among those children everywhere who have difficulty with mathematics.

The erroneous patterns displayed by children are not due to carelessness alone nor are they due to insufficient drill. The patterns are conceptual and they are learned; but that which has been learned is incorrect. Interestingly, although incorrect some procedures produce correct answers once in a while. When this happens, students are reinforced in their belief that they understand what you want them to learn.

You will find that looking for such patterns is a very worthwhile diagnostic activity. By observing such patterns you are able to refrain from assigning drill activities that reinforce incorrect concepts and procedures. Furthermore, you will gain more specific knowledge of each student's strengths and weaknesses upon which to base future instruction; you will save time through more efficient teaching of correct procedures. Your students will benefit more from diagnostic feedback, comments, and precise suggestions than from just having errors marked "wrong."

NEW IN THE SIXTH EDITION

The sixth edition of this text reflects recent research on learning concepts and procedures. For example, both overgeneralization and overspecialization during learning of mathematical concepts and procedures are described and illustrated. New examples of student papers are provided in the current edition, including papers for areas of mathematics other than computation. There is an even greater stress on estimation skills. In regard to instruction, there is increased emphasis on using models; for example, graphic organizers such as cognitive maps and flowcharts. Information on the use of calculators has been expanded, including their use for understanding and for recalling basic facts. There is new material on using cooperative problem solving, and sample tasks are presented which can be given to groups of students. Also, the lists of references for both diagnosis and instruction have been completely revised.

ORGANIZATION OF TEXT

The book is organized into two parts with appendices. Part One sets the stage focusing on several aspects of diagnosis and instruction in computational skills. Helpful guidelines are included. In Part Two, sample student papers are presented. In chapters 3 and 4 you have opportunities to respond to the papers, then feedback is provided in relation to your responses in chapters 4 and 5. Experience has shown that this direct involvement through simulation helps both pre- and inservice teachers become more proficient in diagnosing the specific errors illustrated. You gain skill by actually looking for patterns, making decisions, and planning instruction. At each step, you can "test" your diagnoses and prescriptions for instruction by comparing them with those provided by the author; it is important that you "play the game" and actually take time to respond.

Chapters 3 through 5 can be read in either of two ways: in sequence, or one error pattern at a time as directed in the text. If you choose to focus on one procedure at a time as most readers do, begin your examination of specific error patterns with chapter 3.

Additional student's papers are included in Appendix A where you have many opportunities to practice identifying error patterns; more than 50 student papers are shown. In Appendix B, error patterns are illustrated for geometry, measurement, problem solving, and algebra.

I wish to express appreciation for the encouragement of many classroom teachers who have shown great interest in the material in this book over the years, and acknowledge the help of teachers, former students, and even their students. These colleagues have identified many of the error patterns presented. I would also like to thank the following reviewers who provided valuable comments: Karen A. Verbeke, University of Maryland–Eastern Shore; Michael Schiro, Boston College; Jon M. Engelhardt, University of Texas at El Paso; and Leroy Callahan, The University of Buffalo.

Robert B. Ashlock

Contents

PART
ONE

DIAGNOSIS
and INSTRUCTION

This book is designed to help you become more diagnostically oriented as a teacher. Diagnosis and instruction are both addressed within Part One.

Although the scope of this text is computation (and to some extent other areas of mathematics) many of the ideas presented in Part One can be applied throughout the curriculum by the teacher who wishes to become more sensitive and more responsive to where students are in their development. As students learn concepts and skills in every subject area they often overgeneralize what they are learning, but sometimes they overspecialize instead. Diagnostic teaching involves careful observation. It also attempts to determine what individuals are actually learning—the concepts they are truly learning and the procedures they are really employing.

If you are to teach diagnostically, you must adapt your instruction to what you observe and what you learn about each student. As you learn more about a particular student, tailor your instruction to that child. Then observe and learn more, and adapt your instruction again. Repeat this diagnosis-instruction cycle as often as necessary.

Diagnosing
Error Patterns
in Computation

*Arithmetic is where the answer is right and everything is nice and you can
look out of the window and see the blue sky—or the answer is wrong and
you have to start all over and try again and see how it comes out this time.*[1]

CARL SANDBURG

The child into whose mind Sandburg leads us seems to view arithmetic as an
either-or sort of thing. Either he gets the right answers and he enjoys arithmetic
and life is rosy, or he does not get the right answers and arithmetic and life are
frustrating. You may think this child has a very limited view of arithmetic, and
you may wonder why he is so answer-oriented. Yet, you *do* need to face the
question of why *some* children are not able to get correct answers.

Consider Fred's paper (Figure 1.1). If you merely determine how many
are correct and how many are incorrect, you will not learn *why* his answers are
not correct. Examine Fred's paper and note that when multiplication involves
renaming, his answer is often incorrect. Look for a pattern among the incorrect
responses; observe that he seems to be adding his "crutch" and then multiplying. This can be verified by studying other examples and briefly interviewing
Fred.

You need to examine each child's paper diagnostically; look for patterns,
hypothesize possible causes, and verify your ideas. Approach each child's written work as if each paper is itself a problem or puzzle to be solved.

The diagnosis of errors in arithmetic is an essential part of evaluation in
the mathematics program, but any such diagnosis must be followed by appropriate instruction. It is hoped that reading this book and responding to the
puzzle-like situations presented will help you develop the skills needed to
effectively diagnose errors in computation and provide the instruction required
by each child.

Researchers have known for a long time that we can learn much by carefully observing erroneous procedures. As Hart comments: "We can find out a
considerable amount about a child's stage of understanding if we study his or
her reasoning process when . . . using an erroneous strategy."[2]

3

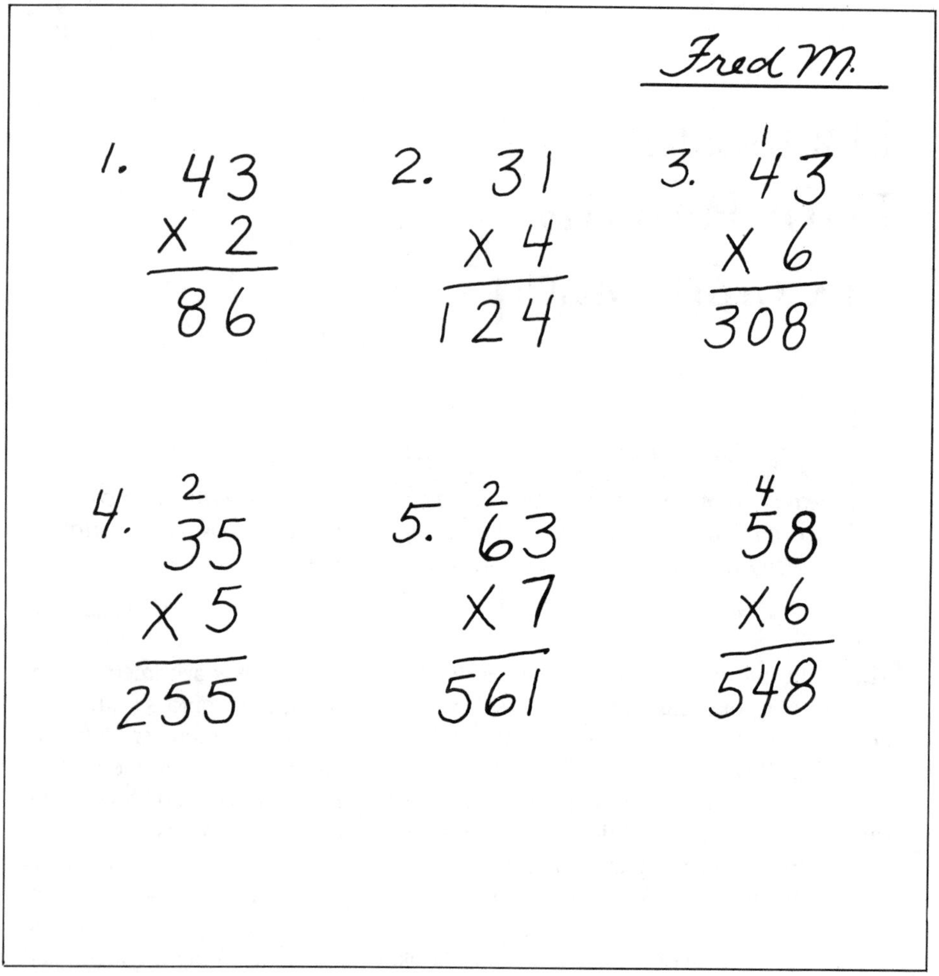

FIGURE 1.1 Sample student paper

COMPUTING WITH ERRORS: Research

Errors in computation are not necessarily just the result of carelessness or not knowing how to proceed. In a study of written computation, Roberts identified four error categories or "failure strategies."[3]

1. *Wrong operation.* The pupil attempts to respond by performing an operation other than the one that is required to solve the problem.
2. *Obvious computational error.* The pupil applies the correct operation, but the response is based on error in recalling basic number facts.
3. *Defective algorithm.*[4] The pupil attempts to apply the correct operation but makes errors other than number fact errors in carrying through the necessary steps.

4. *Random response.* The response shows no discernible relationship to the given problem.

Roberts noted that careless numerical errors and lack of familiarity with the addition and multiplication tables occurred with near-equal frequency at all ability levels. However, using the wrong operation and making random responses were observed more frequently with students of low ability and progressively less frequently with students of higher and higher ability. *The largest number of errors was due to erroneous or incorrect algorithm techniques* in all groups except the lowest quartile, which had more random responses. Incorrect algorithms were used by even the most able achievers (39 percent of the errors made by the upper quartile of pupils studied). Other researchers have made similar observations. Schacht concluded that "differences in performance appear to be of degree and not of kind, with the less able making errors more frequently than the more able."[5] Clearly, the practice papers of *all* pupils must be considered carefully.

In his extension of Roberts' study, Engelhardt classified errors into eight types: basic fact error, defective algorithm, grouping error, inappropriate inversion, incorrect operation, incomplete algorithm, identity error, and zero error.[6] The categories called defective algorithm, grouping error, and inappropriate inversion (all of which involve erroneous procedures) accounted for *61 percent* of the 2279 errors made by third- and sixth-grade children in the study. Again, it is obvious that teachers dare not assume that errors in computation are caused by carelessness or by a child not knowing the basic facts. The actual procedures used are likely to be wrong.

Backman used four categories for procedural errors: errors in sequencing steps within a procedure, errors in selecting information or procedures, errors in recording work, and errors in conceptual understanding.[7]

Brueckner did extensive work identifying types of errors in computation as early as the 1920s. These studies were reported in journals, in his classic 1930 text,[8] and in yearbook articles.[9] He stressed the importance of analyzing written work but also emphasized the need to supplement such activity with interviews. His study of difficulties children have when computing with decimals is characteristic of his research. This report, published in 1928, includes the tabulation of 8785 errors into 114 different kinds.[10] However, many of his categories are ambiguous. Some we would question today because he listed the annexation of unnecessary zeros as an error and considered the vocalization of procedures a faulty habit.

Others have described categories of errors in computation. Guiler noted that in the addition and subtraction of decimals, seven times as many children had trouble with the computation per se than had trouble with the decimal phases of the process. Further, in division of decimals, more than 40 percent of the children placed the decimal point three or more places too far to the right.[11] Such studies remind us of the value of a thorough understanding of numeration and the ability to estimate. In his study of high school students having difficulties with arithmetic, Arthur included types of errors such as adding denominators when adding fractions and failing to invert the divisor when dividing fractions.[12] He concluded that the reteaching of arithmetic skills should be based upon short diagnostic tests and given attention in all high school math classes. Cox concluded from her research that not only did children make systematic errors but, without instructional intervention, they continued with the error patterns for long periods of time.[13] In his study of students in grades 5–8, MacKay reported that many children actually had a high degree of confidence in their erroneous procedures.[14]

Sadowski and McIlveen studied error patterns in sentence-solving.[15] They found that some procedures used by children (although inadequate) produce a correct solution quite frequently, thereby reinforcing continued use of the error pattern. Error studies have also been conducted in algebra. Davis and Cooney studied the work of algebra students and described categories of errors.[16] Carry, Lewis, and Bernard studied errors in solving linear equations and listed types of errors they found.[17] In both studies, patterns of errors were observed.

Noting the similarity of errors across studies, Bright suggested the following categories for future studies of errors in solving linear equations: arithmetic errors, combination errors, transposition errors, operation errors, incomplete solutions, and execution errors.[18]

Several researchers have studied the frequency with which different error patterns occur. One example is Lankford's study of seventh graders.[19] He conducted "diagnostic interviews" of 176 pupils in six schools located in different parts of the United States. The detailed report of the research includes many examples of the use of erroneous algorithms and the frequency of their occurrence among pupils in the study. Lankford's general observations concerning how wrong answers were derived for whole numbers and fractions are included in this text as Appendix C. In his conclusions he notes that "unorthodox strategies were frequently observed—some yielding correct answers and some incorrect ones."[20] Cox's study also gives us insight into the frequency with which erroneous procedures are used.[21] Her study focused on whole number algorithms and compared the work of children in "regular" classrooms with computation done by children placed in special education classrooms. A third study, conducted by Research for Better Schools, Inc., and reported by Graeber and Wallace, examined addition, subtraction, and multiplication of whole numbers on individually prescribed instruction (IPI) mathematics pretests.[22] In both the Cox study and the IPI study, an erroneous procedure was not classified as "systematic" unless it occurred at least three times.

As you examine a student's paper, remember that at least 20% of arithmetic errors are likely to be careless errors.[23] In view of this, always use more than one question when you diagnose a particular concept or skill. Use multiple questions so you can observe patterns.

Resnick distinguishes between what she calls the syntax and the semantics of a computational procedure.[24] Syntax includes the arrangement of symbols and the mechanics of procedural rules of the algorithm, whereas semantics refers to the mathematical meanings or principles involved. She believes that students who develop error patterns often focus on the syntax. They often manipulate digits and ignore the quantities involved.

> Systematic errors probably arise from a basic failure to mentally represent arithmetic procedures in terms of operations on quantities within a principled number system, rather than as operations on symbols that obey largely syntactic rules.[25]

Resnick also argues that relatively few conceptual misunderstandings are at the heart of many error patterns.[26]

A recent research and development project at the University of Maryland was concerned with misconceptions of secondary school students and patterns of erroneous processes common among such students. The investigators classified the misconceptions they observed into four categories: overgeneralizations, overspecializations, mistranslations, and limited conceptions or "fragile knowledge."[27] They found that many of the misconceptions had logical explanations but often inadequate frames of reference. However, they also learned that many experienced teachers did not acknowledge that there were logical explanations.

For a long time teachers have known that many of the errors students make are error patterns or systematic errors. In more recent years the similarity of these errors to bugs in a computer program has been noted and, as a result, systematic errors are often called *bugs* in professional literature.

Attempts have been made to construct computer programs for diagnosing systematic student errors. One such attempt, DEBUGGY, was found to be a valuable research tool, although some error patterns were not diagnosed. Repair Theory was developed to explain many of the remaining patterns.[28] While describing his work with Repair Theory, VanLehn reported that the bugs or error patterns of some students are unstable from one test to the next. It appeared that error patterns are sometimes used as problem-solving strategies; they are used by a student only long enough to get the student through the test.[29] Of course, other error patterns are used tenaciously, even during instruction. The potential of computer programs to help classroom teachers diagnose systematic student errors remains unclear.

LEARNING ERROR PATTERNS

How do children learn error patterns for computation? Children's mathematical ideas and computational procedures may be correct or erroneous, but the *process* of abstracting those ideas and procedures is basically the same. From a set of experiences with a concept or a process, a child pulls out or abstracts those things which the experiences have in common. The intersection of the experiences defines the idea or process for the child.

If a child's only experiences with the idea *five* are with manila cards having black dots in the familiar domino pattern (Figure 1.2a) he may abstract from these experiences a notion of five that includes many or all of the characteristics his experiences have had in common. The ideas of black on manila paper, round dots, or a specific configuration may become part of the child's idea of five. One

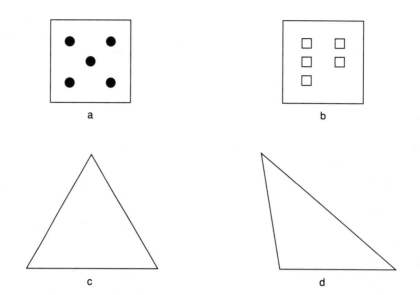

FIGURE 1.2 Pattern boards and triangles

of the author's students, when presenting to children the configuration associated with Stern pattern boards (Figure 1.2b) was told, "That's not five. Five doesn't look like that."

More children will name as a triangle the shape in Figure 1.2c than the shape in Figure 1.2d; yet both are triangles. Again, configuration (or even the orientation of the figure) may be a common characteristic of the child's limited range of experiences with triangles.

Dr. Geoffrey Matthews, organizer of the Nuffield Mathematics Teaching Project in England, told about a child who computed correctly one year but missed half of the problems the next year. As the child learned to compute, he adopted the rule, "Begin on the side by the piano." The next school year the child was in a room with the piano on the other side, and he was understandably confused about where to start computing.

When multiplication of fractions (or rational numbers) is introduced, children frequently have difficulty believing their correct answers make sense because throughout their previous experiences with factors and products, the product was always at least as great as the smaller factor. (Actually, the product is noticeably greater than either factor in most cases.) In the mind of the child, the concept of product had come to include the idea of greater number, because this was common throughout most of his experiences with products. This may be one reason many children have difficulty with the zero property for multiplication of whole numbers. The product zero is usually less than one of the factors.

Creature Cards also illustrate this view of concept formation (Figure 1.3). The child confronted with a Creature Card is given a name or label such as Gruffle and told to decide what a Gruffle is. He looks for common characteristics

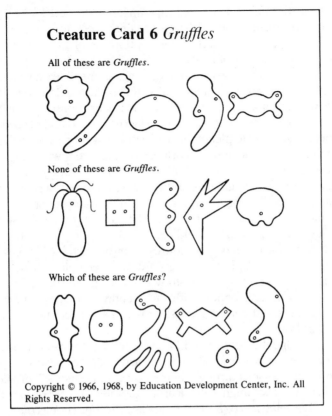

FIGURE 1.3 Creature card (*Source: Reproduced from the Elementary Science Study unit,* Attribute Games and Problems, *by permission of Education Development Center, Inc.*)

among a set of Gruffles. Then, experiences with non-Gruffles help him eliminate from consideration those characteristics which happen to be common in the set of Gruffles but are not essential to "Gruffleness" (i.e., the definition of a Gruffle would not include such attributes). Finally, the cards provide an opportunity for the child to test out his newly derived definition.

Children often learn erroneous concepts and processes similarly. They look for commonalities among their initial contacts with the idea or procedure. They pull out or abstract certain common characteristics, and their concept or algorithm is formed. The common attributes may be very specific, such as crossing out a digit, placing a digit in front of another, or finding the difference between two one-digit numbers (regardless of order). Failure to consider enough examples is one of the errors in inductive learning often cited by those who study thinking and learning.[30]

From time to time such inadequate procedures produce correct answers. When this happens, use of the erroneous procedure is reinforced for the child who is anxious to succeed. The child who decides that rounding whole numbers to the nearest ten means erasing the units digit and writing a zero is correct about half of the time! After observing similar "discoveries" by children, the psychologist Friedlander noted that as teachers "seek to capitalize on the students' reasoning, errors of fact, of perception, or of association can lead to hopelessly chaotic chains of mistaken inferences and deductions."[31]

There are many reasons why children are prone to learn patterns of error. It most certainly is not the intentional result of instruction. Yet all too often, especially when taught in groups, children do not have prerequisite understandings and skills they need when introduced to new ideas and procedures. When this happens, they want to please the teacher (or at least survive in the situation), so they tend to "grab at straws." Furthermore, teachers who introduce paper-and-pencil procedures while a child still needs to work problems out with concrete aids are encouraging the child to try to memorize a complex sequence of mechanical acts. This, again, prompts the child to adopt simplistic procedures he can remember. Because incorrect algorithms do not usually result in correct answers, it would appear that a child receives limited positive reinforcement for continued use of erroneous procedures. However, children sometimes hold tenaciously to incorrect procedures even during remedial instruction. Each incorrect algorithm is an interesting study in itself, and in succeeding sections of this book you will have opportunities to identify several erroneous computational procedures and consider possible reasons for adoption of such procedures by children.

Keep in mind the fact that children who learn patterns of error *are* capable of learning. Maurer even claims they are "acting like creative young scientists, interpreting their lessons through their own generalizations."[32] Typically, these children have what we might call a learn*ed* disability, not a learn*ing* disability.[33] In *Children's Arithmetic: The Learning Process,* Ginsburg points out that the rules young children adopt are meaningful to them.[34] Their rules are derived from a search for meaning, and a sensible learning process is involved. This is true even for the erroneous rules they invent, though such rules may involve a distortion or a poor application. More than computational procedure is in view here. For example, Ginsburg describes the invention of erroneous rules as children learn to count.[35] Children can also be observed inventing similar rules when introduced to the equals sign, "The equals sign means 'the answer turns out to be.' "[36]

OVERGENERALIZING AND OVERSPECIALIZING

Many erroneous concepts and procedures are generated when a child overgeneralizes during the learning process. Other error patterns may be invented when a child overspecializes.

We are all prone to overgeneralize; we "jump to a conclusion" before we have adequate data at hand. Examples of overgeneralizing abound in many areas of mathematics learning. Several interesting examples were observed by project staff at the University of Maryland during their study of misconceptions among secondary school students.[37]

What is a sum? Sometimes children decide that a sum is the number that is written on the right side of an equals sign.

$$4 + 2 = 6$$

Both are considered sums.

$$6 - 2 = 4$$

Consider students who agree that all three of these figures are triangles.

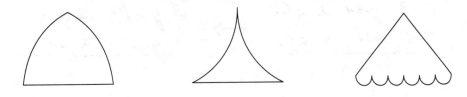

Graeber reports an interesting speculation on this situation.

These students may be reasoning from a definition of triangle position. Extension of this definition to simple closed curves that are not polygons may lead to this error of including such shapes in the set of triangles.[38]

Sometimes children are exposed to right triangles such as these.

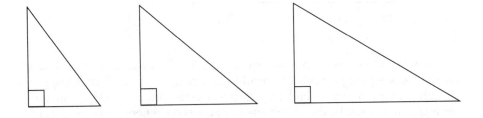

Children conclude that in addition to measuring 90 degrees, a right angle is oriented to the right.

a right angle   . . . therefore . . .   a left angle

The child who believes that 2y means 20 + y may be overgeneralizing from expressions like 23 = 20 + 3. Other children always use ten for regrouping, even when computing with measurements.

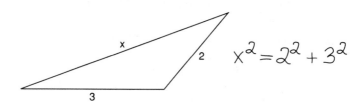

Secondary school students sometimes think of the longest side of a triangle as a hypotenuse. They assume the Pythagorean Theorem applies even when the triangle is not a right triangle.[39]

$$x^2 = 2^2 + 3^2$$

Other erroneous procedures are generated when a child *overspecializes* during the learning process. The resulting procedures are restricted inappropriately. For example, a child may decide that in order to add or subtract decimals, there must be the same number of digits on either side of the decimal point. Therefore, a child may rewrite 100.36 + 12.57 as 100.36 + 125.70. Also, children know that in order to add or subtract fractions, the fractions must have like denominators. Sometimes children believe that multiplication and division of fractions requires like denominators.[40]

It is quite common for students to restrict their concept of altitude of a triangle to only that which can be contained within the triangle.[41]

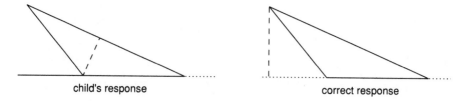

child's response   correct response

As you diagnose children experiencing difficulty, be alert for both over-generalization and overspecialization. Probe deeply and try to learn why specific procedures were acquired. Your discoveries of overgeneralizations and over-specializations will help you provide needed instruction.

DIAGNOSING WITH TESTS

As you diagnose a child's concepts and skills in computation, much of what you learn comes from paper-and-pencil tests or other written work. In a sense, this entire book is designed to help you learn as much as possible from the written work of children.

For diagnostic purposes, standardized achievement tests have only limited value. They usually sample such a broad range of content that you do not learn what you need to know about specific concept and skill categories. According to Kamii and Lewis, achievement tests emphasize lower-order thinking and can result in misleading information.[42] On the other hand, diagnostic tests sample a narrower range of content, often in a way that permits you to identify areas of strength and weakness. The KeyMath-R is an example of a commercially available diagnostic test.

KeyMath-R[43]

This test is intended for use with children in kindergarten through grade 9. It is administered to children individually, and its exceptionally attractive format maintains a child's interest during that administration. Thirteen untimed subtests are included, and two forms of the test are available for pre- and posttesting.

KeyMath-R is a "diagnostic inventory of essential mathematics." It focuses significantly on operations and on algorithms for the operations with whole and rational numbers. Mental computation and estimation are also included, as are geometry and problem solving (both routine and non-routine). Some NCTM curriculum standards are not well represented: communication (other than interpreting graphs, charts, and tables), reasoning, patterns and relationships, and mathematical connections.

KeyMath Teach and Practice[44]

These materials are an attempt to reflect the NCTM curriculum standards more completely than the KeyMath-R, in a form that can be used for diagnosis and instruction within a classroom setting. The probes, worksheets, and lesson cards focus on basic concepts, operations and applications—including mental computation, interpreting data, and problem solving.

An extensive listing of mathematics tests available in the United States and Canada can be obtained from the National Council of Teachers of Mathematics.[45]

Computers appear to have great potential as tools to assist with diagnosis of computational skills. Much of the software that is currently available is not greatly advanced beyond a set of paper-and-pencil tests, so make sure that use of the computer program is preferable to a paper-and-pencil procedure.

Most teachers do not have time to interview *each* child. From time to time you probably need to prepare your own diagnostic test for a specific concept or skill category. You may need a test to administer to children who have difficulty subtracting when regrouping is involved. The error patterns presented in Chapter 3 and discussed in Chapter 4 suggest distractors you can use for test items. Distractors drawn from error patterns may give you clues to the child's thought process. The following multiple-choice item was built from error patterns; each distractor is an answer a child might choose if the child has learned an erroneous procedure.

The answer is:

4372	a. 2526
− 2858	b. 1514
	c. 524
	d. 2524

If you need to prepare a focused diagnostic test, you may want to base a few items on cognitive maps or flowcharts with which students have worked during instruction. Figure 1.4 is an example of a test item using a concept map,

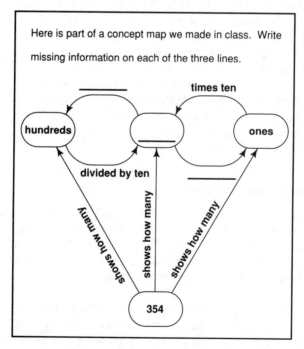

FIGURE 1.4 A teacher-made test item using a concept map

and Figure 1.5 is a test item which utilizes a flowchart. In both items the child is to write appropriate words in the designated spaces.

INTERVIEWING AND OBSERVING

Written tests are helpful in diagnosis, but they are limited. Interviews have long been recognized as an effective way to collect additional information about a child's mathematical concepts, to gain both quantitative and qualitative data. Interviewing is not just "oral testing" to determine whether a child can do a task. In an interview you can focus on the child's thinking and determine if that child is learning in a meaningful way. You can determine what strategies are used to solve problems.

An interview is not a time for expressing your opinion or asking questions that are merely prompted by your own curiosity. Rather, it is a time to observe the child carefully and a time to *listen*. Adapt the pace of the interview so the child can respond comfortably. Be careful not to give clues or ask leading questions. It has been said that we are all born with two ears and one mouth, and we probably should use them in that proportion. This applies quite specifically to teachers, who sometimes want to talk and explain when they should listen.

Record the child's responses as you proceed with the interview. Make written notes or tape record the interview and write down only those things that will not appear on tape, say, an expression of delight or disgust. You may also want to record your judgments as to the child's level of understanding. Do not rely on your memory to make a true record.

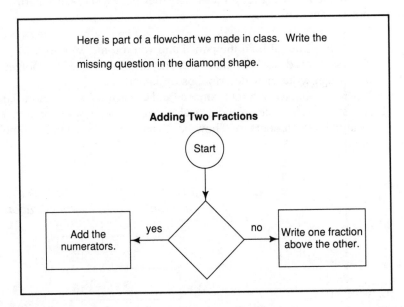

FIGURE 1.5 A teacher-made test item using a flowchart

Interviews can vary widely in their structure. You may ask a child to choose among the alternatives you present, but generally you will want the child to respond freely. To get at a child's thinking, you will want the child to comment on his or her own thought processes. This can be accomplished either through introspection or through retrospection. To elicit introspection, ask the child to comment on thoughts as the task is being done; have the child "think out loud." Of course, the very process of commenting aloud can influence the thinking a child does.[46] To elicit retrospection, ask the child to comment on thoughts after a task is completed. Remember that, when commenting in retrospect, a child may forget wrong turns that were taken, etc. It is probably best to elicit introspection part of the time and retrospection part of the time.

Stimulus situations should be given in varied modes. Consider this:

Mode	*Examples*
Verbal	Words, oral and written
Written symbols	Numerals
Two-dimensional representations	Paper-and-pencil diagrams, photographs
Three-dimensional representations	Base ten blocks, math balance, place value chart

A stimulus situation can be given in any of the above modes. Each stimulus situation will call for a response, and responses for a given stimulus can be similarly varied. If a child's responses are ambiguous, vary the questions in order to draw out the child.

Consider the following techniques; they may help you obtain information you will not get any other way.

1. Say, "This time I'll hold the pencil and you tell me what to do."
2. Have students describe to other students how they solved a problem, or have them write their descriptions on paper.
3. Provide a slightly different context, and ask students to use the idea.
4. Ask students to rate mathematical examples on a scale of 1–4. For example, ask students to "Show how you see each example as multiplication."

	It Is Multiplication				*It Is Not Multiplication*
A. 5×3	1	2	3	4	
B. $(-5)(-3)$	1	2	3	4	
C. $\sqrt{5}\pi$	1	2	3	4	

5. Discuss the responses with the child. Children who interpret multiplication only as repeated addition usually consider example *A* to be more truly an example of multiplication than *B* or *C*.

During the interview carefully observe students' behavior in response to their own errors. Be sure to diagnose each student's ability to estimate; instruction may well need to focus on this important skill and the concepts involved. Remember that students with learning disabilities often communicate information that is incorrect, yet it is what they actually see.

Also look carefully at the child's understanding of the equivalence relationship. What does an equals sign really mean to the child? It may mean "results in" rather than "is the same as." Mevarech and Yitschak note that "... children at various age levels interpret the symbol of equivalence, the equal sign, as an operation rather than a relation symbol."[47]

INTERVIEWING AND REFLECTING

In recent years researchers have become increasingly interested in metacognition, that is, in what a child knows about his or her own cognitive performance and the child's ability to regulate that performance. Garofalo maintains that "mathematics educators are convinced that what one knows or believes about oneself as a learner and doer of mathematics and how one controls and regulates one's behaviors while working through mathematical tasks can have powerful effects on one's performance."[48]

Many of the questions you ask while interviewing a child will help the child become more aware of his own cognitive processes. For example:

How did you get that answer?
If you had to teach your brother to do this, how would you do it? What would you say to him?

Garofalo lists the following sample questions.[49]

Think of everything you do when you practice solving mathematics problems. Why do you do these things?
What kinds of errors do you usually make? Why do you think you make these errors? What can you do about them?
Tell why taking one's time might be important. Is it always important?
What do you do when you see an unfamiliar problem? Why?
Name some things that you can do when you are stuck on a problem. Do these things always help?
Name several things that you can do to keep track of what you're doing.
Tell when it's useful to check your work. Why? Is that the only time?
What kinds of things do you forget to do when solving mathematics problems?
What kinds of problems are you best at? Why?

What kinds of problems are you worst at? Why? What can you do to get better at these?

Following is a transcript of part of an interview I had with a fourth grader. We were discussing what she had written:

$$\frac{1}{3} = \frac{4}{12}$$
$$+$$
$$\frac{1}{4} = \frac{3}{12} \quad \frac{7}{12}$$

TEACHER: What do the equal signs tell us?
STUDENT: They tell us . . . that you just do the answer.
TEACHER: Which is more, one third or four twelfths?
STUDENT: (pause) one third? . . . no
TEACHER: What does equals mean?
STUDENT: (no response)

This child knows a procedure, but her understanding of concepts is inadequate. The interview may help her evaluate her own thinking.

While a student is solving a verbal problem, determine how the student represents the problem mentally. Does the student conceive the problem in terms of a whole and its parts? Can the situation be shown with a diagram or manipulatives? It is often helpful to have students retell a problem in their own words.

More informal observation is also a source of information about a child, whether in the classroom or on the playground. Watch students as they play mathematical games. Note how they get the information they need. Intervene with diagnostic questions from time to time.

Some instructional activities help students monitor their own cognitive activity and at the same time provide diagnostic information for the teacher. Have students keep a journal and write spontaneously about any of the following.

Describe processes.
Summarize what was learned.
Describe what they expect to happen.
Describe how they felt during a lesson.

You may want students to keep two journals: one to write what they know or have learned and another to tell how they feel about specific experiences in mathematics. As Rin notes, when students write mathematical prose they often expose inadequate understandings of definitions and incorrect usage of terms and symbols.[50]

You may also want to have students communicate what they know about mathematics by having them make concept maps. Concepts are written on paper, and then relationships are shown with lines. Linking words (usually verbs) can also be added. Emphasize to students the importance of putting *their* thoughts on paper.[51] Figure 1.6 illustrates how two fourth-grade boys in the same classroom drew very different maps for the word "subtraction." Figure 1.7 shows how two fourth graders responded to "fractions." Note that one child associated fractions with drawings, which she labeled incorrectly.

GUIDING DIAGNOSIS

A child's work must not only be scored, it must be analyzed as well if it is to provide useful information for diagnosis. Many times it is possible for students to mark which examples are correct or incorrect. Then you can spend more of your professional time analyzing the written work of children and planning needed instruction.

Observe what the child does and does not do; note the computation that has a correct answer and the computation that does not; and look for those procedures used by the child that might be called mature and those that are less mature. Distinguish between situations in which the child uses an incorrect procedure and situations in which the child does not know how to proceed at all.

The following statements provide a brief summary of principles to keep in mind as you diagnose the work of children having difficulty with computation.

1. *Be accepting.* Diagnosis is a highly personal process. Before a child will cooperate with you as his teacher in a manner that may lead to lessening or elimination of his problems with computation, he must perceive that you are interested in and respect him as a person, that you are genuinely interested in helping him, and that you are quite willing to accept a response even when the response is not correct. In short, if sufficient data are to be collected for an adequate diagnosis, the child must understand that you are willing to accept his failures. You must exhibit something of the attitude of a good physician toward his patient. As Tournier, a Swiss physician, has noted, "What antagonizes a patient is not the truth, but the tone of scorn, pity, criticism, or reproof which so often colors the statement of the truth by those around him."[52]

2. *Collect data—do not instruct.* In making a diagnosis, you must differentiate between the role of collecting data (testing) and the role of teaching. Diagnosing involves gathering as much useful data as possible and making judgments on the basis of the data collected; in general, the more data, the more adequate the judgments which follow. The child is apt to provide many samples of incorrect and immature procedures if he sees that you are merely collecting information that will be used to help him overcome his difficulties. However, if you point out errors, label responses as "wrong," and offer instruction, the child is far less likely to expose his own inadequate performance. When a teacher who usually offers help as soon as he sees incorrect or immature performance begins to

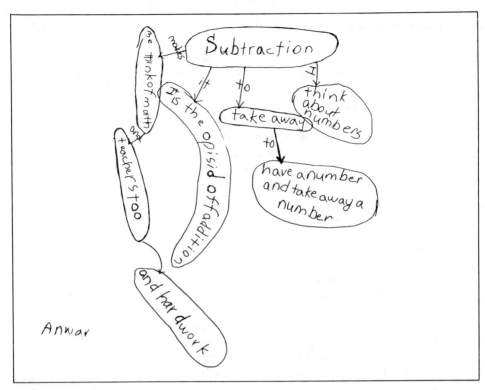

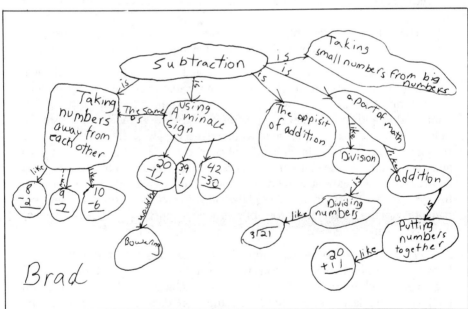

FIGURE 1.6 Cognitive maps for subtraction by fourth-grade boys

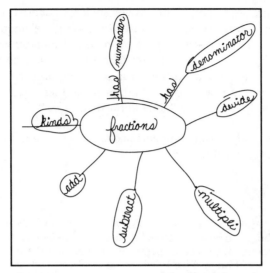

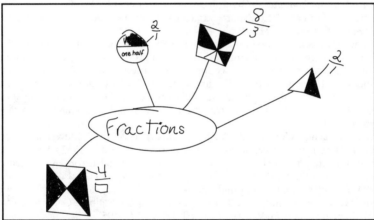

FIGURE 1.7 Cognitive maps for fractions by fourth graders

distinguish between collecting data and instruction, he is often delighted with the way children begin to open up and lay bare their thinking.

3. *Be thorough.* A single diagnosis is rarely thorough enough to provide direction for continuing instruction. From time to time it is necessary to have short periods of diagnosis even during instruction. In fact, if you are alert during the instruction that follows a diagnosis, you may pick up cues that suggest additional diagnostic activities apt to provide helpful data about the child and his problem. Whenever possible say, "Can you read it another way?" or "Can you do it another way?" Use the child's own language when possible, but not necessarily the vernacular of the day. Allow children plenty of thinking time, and observe their behavior.

4. *Look for patterns.* Data should be evaluated in terms of patterns, not isolated events. A decision about remedial instruction can hardly be based upon col-

lected bits of unrelated information. As you look for patterns you engage in a kind of problem-solving activity, because you look for elements common to several examples of a child's work. Look for repeated applications of erroneous definitions and try to find consistent use of incorrect or immature procedures. The importance of looking for patterns can hardly be overstressed. Many erroneous procedures are practiced by children while their teachers assume the children are merely careless or "don't know their facts."

For further study, serious students of diagnosis will want to examine many of the references listed after Chapter 5.

ENDNOTES

1. Excerpt from "Arithmetic" in *The Complete Poems of Carl Sandburg*, copyright 1950 by Carl Sandburg and renewed 1978 by Margaret Sandburg, Helga Sandburg Crile and Janet Sandburg. Reprinted by permission of Harcourt Brace Jovanovich, Inc.
2. Kathleen M. Hart, "I Know What I Believe; Do I Believe What I Know?" *Journal for Research in Mathematics Education* 14, no. 2 (March 1983): 120.
3. Gerhard H. Roberts, "The Failure Strategies of Third Grade Arithmetic Pupils," *The Arithmetic Teacher* 15 (May 1968):442–46.
4. An algorithm is a step-by-step written procedure for determining the result of an arithmetic operation (*i.e.*, a sum, a difference, a product, or a quotient). A variety of algorithms or computational procedures can be used to determine any one missing number; and indeed, at different times and in different parts of the world, many differing but useful algorithms are taught. However, the algorithm used by a child is said to be "incorrect" or "defective" if the procedure does not always produce the correct result.
5. Elmer J. Schacht, "A Study of the Mathematical Errors of Low Achievers in Elementary School Mathematics," *Dissertation Abstracts* 28A (September 1967): 920–21.
6. Jon M. Engelhardt, "Analysis of Children's Computational Errors: A Qualitative Approach," *British Journal of Educational Psychology* 47 (1977): 149–54.
7. Carl A. Backman, "Analyzing Children's Work Procedures," in *Developing Computational Skills*, Marilyn Suydam, ed., 1978 yearbook of the National Council of Teachers of Mathematics (Reston, Va.: The Council, 1978), 177–195.
8. Leo J. Brueckner, *Diagnostic and Remedial Teaching in Arithmetic* (Philadelphia: John C. Winston Co., 1930).
9. For example, his "Diagnosis in Arithmetic," in *Educational Diagnosis*, 34th Yearbook, National Society for the Study of Education (Bloomington, Ill.: Public School Publishing Co., 1935), 269–302.
10. Leo J. Brueckner, "Analysis of Difficulties in Decimals," *Elementary School Journal* 29 (September 1928): 32–41.
11. Walter S. Guiler, "Difficulties in Decimals Encountered in Ninth-Grade Pupils," *Elementary School Journal* 46 (March 1946): 384–93.
12. Lee E. Arthur, "Diagnosis of Disabilities in Arithmetic Essentials," *The Mathematics Teacher* 43 (May 1950): 197–202.
13. L. S. Cox, "Diagnosing and Remediating Systematic Errors in Addition and Subtraction Computations," *The Arithmetic Teacher* 22 (February 1975): 151–57.
14. I. D. MacKay, *A Comparison of Students' Achievement in Arithmetic with Their Algorithmic Confidence* (Vancouver, B.C.: Mathematics Education Diagnostic and Instructional Centre, University of British Columbia, 1975.) (ERIC Document Reproduction Service No. ED 128 228)
15. Barbara R. Sadowski and Delayne Houston McIlveen, "Diagnosis and Remediation of

Sentence-solving Error Patterns," *The Arithmetic Teacher* 31 (January 1984): 42–45.

16. Edward J. Davis and Thomas J. Cooney, "Identifying Errors in Solving Certain Linear Equations," *The MATYC Journal* 11 (1977): 170–78.

17. L. Ray Carry et al., *Psychology of Equation Solving: An Information Processing Study, Final Technical Report* (Austin, Texas: Department of Curriculum and Instruction, University of Texas, 1979). (ERIC Document Reproduction Service No. ED 186 243)

18. George W. Bright, "Student Errors in Solving Linear Equations," *RCDPM Newsletter* 6, no. 2 (Fall 1981): 3–4.

19. Francis G. Lankford, Jr., *Some Computational Strategies of Seventh Grade Pupils*, U.S. Department of Health, Education, and Welfare, Office of Education, National Center for Educational Research and Development (Regional Research Program) and The Center for Advanced Study, The University of Virginia, October 1972. (Project number 2-C-013, Grant number OEG-3, 72-0035)

20. *Ibid.*, 40.

21. L. S. Cox, "Systematic Errors in the Four Vertical Algorithms in Normal and Handicapped Populations," *Journal for Research in Mathematics Education* 6, no. 4 (November 1975): 202–20.

22. Anna O. Graeber and Lisa Wallace, *Identification of Systematic Errors: Final Report* (Philadelphia: Research for Better Schools, Inc., 1977). (ERIC Document Reproduction Service No. ED 139 662).

23. See M.A. Clements, "Careless Errors Made by Sixth-Grade Children on Written Mathematical Tasks," *Journal for Research in Mathematics Education*, (March 1982): 136–144.

24. Lauren B. Resnick, "Beyond Error Analysis: The Role of Understanding in Elementary School Arithmetic," in *Diagnostic and Prescriptive Mathematics: Issues, Ideas, and Insights*, Helen N. Cheek, ed., 1984 research monograph of the Research Council for Diagnostic and Prescriptive Mathematics (Kent, Ohio: The Council, 1984), 2–14.

25. *Ibid.*, 13.

26. *Ibid.*, 2.

27. Anna O. Graeber, principal investigator, *Methods and Materials for Preservice Teacher Education in Diagnostic and Prescriptive Teaching of Secondary Mathematics: Project Final Report* (University of Maryland, January 1992), NSF funded grant.

28. John S. Brown and Kurt VanLehn, "Repair Theory: A Generative Theory of Bugs in Procedural Skills," *Cognitive Science* 4 (1980): 379–426.

29. Kurt VanLehn, *Bugs Are Not Enough: Empirical Studies of Bugs, Impasses and Repairs in Procedural Skills* (Palo Alto, Calif.: Cognitive and Instructional Sciences Group, Xerox, Palo Alto Research Center, March 1981).

30. For example, see R. S. Nickerson, D. N. Perkins, and E. E. Smith, *The Teaching of Thinking* (Hillsdale, N.J.: Lawrence Erlbaum, 1985), 111.

31. Bernard Z. Friedlander, "A Psychologist's Second Thoughts on Concepts, Curiosity, and Discovery in Teaching and Learning," *Harvard Educational Review* 35 (Winter 1965): 29.

32. Stephen B. Maurer, "New Knowledge about Errors and New Views about Learners: What They Mean to Educators and More Educators Would Like to Know," in *Cognitive Science and Mathematics Education*, Alan H. Schoenfeld, ed., (Hillsdale, N.J.: Lawrence Erlbaum, 1987), 165–187.

33. These terms were used frequently by John Wilson at the University of Maryland before his death.

34. Herbert Ginsburg, *Children's Arithmetic: The Learning Process* (New York: D. Van Nostrand Company, 1977). See especially Chapter 6.

35. Ginsburg, *op. cit.*, Chapter 1.

36. See Merlyn Behr, Stanley Erlwanger, and Eugene Nichols, "How Children View the Equals Sign," *Mathematics Teaching* 92 (September 1980): 13–15, or a paper with the same title available from ERIC Document Reproduction Service No. ED 144 802.

37. See Anna O. Graeber, principal investigator, *Methods and Materials for Preservice*

Teacher Education in Diagnostic and Prescriptive Teaching of Secondary Mathematics: Project Final Report (University of Maryland, January 1992), NSF funded grant, Chapter 4, pp. 4–49.

38. *Ibid.*, 4–5.
39. *Ibid.*, 4–12.
40. *Ibid.*, 4–35.
41. *Ibid.*, 4–31.
42. Constance Kamii and Barbara Ann Lewis, "Achievement Tests in Primary Mathematics: Perpetuating Lower-Order Thinking," *The Arithmetic Teacher* 38 (May 1991): 4–9.
43. American Guidance Service, 1988.
44. American Guidance Service, 1992.
45. James S. Braswell and Alicia A. Dodd, *Mathematics Tests Available in the United States and Canada* (Reston, Virginia: NCTM, 1988).
46. See Alan H. Schoenfeld, "Making Sense of 'Out Loud' Problem-Solving Protocols." *The Journal of Mathematical Behavior* 4 (October 1985): 171–191.
47. Zemira R. Mevarech and Dostis Yitschak, "Students' Misconceptions of the Equivalence Relationship," in *Proceedings of the Seventh International Conference for the Psychology of Mathematics Education*, Rina Herschkowitz, ed. (Rehovat, Israel: Weizman Institute of Science, 1983): 313.
48. Joe Garofalo, "Metacognition and School Mathematics," *The Arithmetic Teacher* 34 (May 1987): 22.
49. *Ibid.*, 22–23.
50. Hadas Rin, "Linguistic Barriers to Students' Understanding of Definitions," in *Proceedings of the Seventh International Conference for the Psychology of Mathematics Education*, Rina Herschkowitz, ed. (Rehovat, Israel: Weizman Institute of Science, 1983): 295–300.
51. For suggestions on introducing concept maps to children, see Joseph D. Novak and D. Bob Gowin, "How to Introduce Students to Concept Mapping." *Learning How to Learn* (New York: Cambridge University Press, 1984): 24–39.
52. Paul Tournier, *The Healing of Persons* (New York: Harper and Row, 1965): 243.

Providing Needed Instruction

You know that when Fred multiplies he usually adds the "crutch" before multiplying, so you need to plan instruction for him. This chapter is designed to help you provide effective instruction. Later, in Chapter 5, specific suggestions are listed for particular error patterns.

When teaching computational procedures, interweave instruction with diagnosis. Be alert to what each child is actually doing and eager to probe deeper. Be willing to change your plans as soon as what you see or hear suggests that an alternative would be more fruitful in the long run. Diagnostic teaching is, first of all, an attitude that cares very much about each child's learning.

If you think about it for a moment you will realize that diagnostic teaching is cyclical. After your initial diagnosis, you plan and conduct the lesson, but what you see and hear during the lesson prompts you to modify your previous judgment and possibly seek more information before planning the next lesson. Sometimes you move through the cycle very rapidly several times in the course of a single lesson. At other times, it occurs over a span of several lessons.

As you plan instruction, keep in mind the teaching sequences that are normally followed when algorithms are first introduced. An overview of instruction is presented in this chapter, but you may want to consult a methods text for a more complete discussion.

As you read the following sections, note the stress on estimation and the importance of using manipulatives appropriately. You will be encouraged to make sure children understand numerals sufficiently before teaching them to compute, and also to teach students appropriate use of calculators in relation to paper-and-pencil computation. Guidelines for instruction focus on the child, underlying concepts and skills, and skill in computational procedures.

UNDERSTANDING CONCEPTS AND PRINCIPLES

One of the chief reasons children have difficulty learning to compute is they do not have sufficient understanding of the concepts and principles that underlie the algorithms. When children are introduced to whole number algorithms, their understanding of multi-digit numerals is not always adquate to provide a base for learning computation procedures which make sense to them.

Numeration

Understanding our Hindu-Arabic numerals is *not* just "knowing place value." The concept of place value is important, but it is but one of many ideas children need to know if they are to understand multi-digit numerals and learn computational procedures readily. Consider this principle: "A multi-digit numeral names a number which is the *sum* of the products of each digit's face value and place value."[1] The terms used in this definition alert us to a number of ideas that are incorporated in an understanding of multi-digit numerals.

In order to understand multi-digit numerals, a child must first have some understanding of the operations of addition and multiplication. A child must also be able to distinguish between a digit and the complete numeral. An understanding of a digit's face value involves the cardinality of the numbers zero through nine. The idea of place value involves the assignment of a value to each position within a multi-digit numeral; that is, each place within the numeral is assigned a power of ten. We, therefore, identify and name the tens place and the thousands place. This rather specific association of value with place is independent of whatever digit may happen to occupy the position within a given numeral.

Usually, children having difficulty can identify and name place values, but they cannot get the next step. They have not learned to combine the concepts of face value and place value. It is the *product* of a digit's face value and its place value, sometimes called "total value of the digit" or "product value," which must be used. The sum of such products is the value of the numeral, and, in renaming a number, these products of face value and place value must continually be considered.

Ginsburg observes that "it takes several years for children to master the place value system for writing numbers."[2] In fact, in the very first of the three stages he proposes with reference to a child's understanding of numerals, the child can already write the numeral correctly. However, the child cannot explain why it is written that way.[3]

When teaching children our numeration system, we should introduce numerals as a written record of observations made while looking at or manipulating objects. For multi-digit numerals, these observations frequently follow manipulating the materials according to accepted rules in order to obtain the fewest pieces of wood (or the like), and, thereby, representations for the standard or simplest name for a number are obtained.

As Resnick notes, "Conceptual understanding does not always automatically produce correct procedures ... instruction also must address directly the problem of linking conceptual understanding to procedural skill."[4] As we teach children, we need to make the link between principles and written notation as explicit as possible.

In all activities where children associate a numeral with materials, it is important that they have opportunities to go both ways. Children may be given materials to sort, regroup, trade, etc., and then they record the numeral that shows how much is observed. At the same time, children need to be given multi-digit numerals to interpret by constructing a set of materials that shows how much the numeral represents. The ability of a child to go from objects to symbol and also from symbol to objects is an important indicator that the child is coming to understand the meaning of multi-digit numerals.

There are many aids available for numeration instruction, most of which are also used for demonstrating computational procedures. If teachers consider these aids carefully, they will recognize that it is possible for some aids to be used with little or no understanding of concepts such as face value, place value, and product value. Frequently, children merely learn complex mechanical procedures for getting answers.

Involve children with materials that make it possible to compare the value of a collection of objects with the equivalent value of a single object (e.g., base blocks). Later, use aids in which many objects are traded for a single object that is identical to those objects except for its placement (e.g., sticks in place-value cans, or trading activities with chips of one color on a place-value mat). These aids are helpful because they more accurately picture the way digits are used within multi-digit numerals.

Other Concepts and Principles

Numeration concepts and principles are related to other concepts and principles incorporated within computational procedures. In addition to place value (the principle that the values which digits represent are determined by the column in which a digit is written), Resnick lists four principles incorporated in subtraction algorithms:[5]

1. Difference principles. These state the relationship between the two numbers in the example.
2. Composition principle. Each quantity can be expressed as a composition of lesser quantities; e.g., a two-digit number can be expressed as tens and units.
3. Partition principle. Quantities can be partitioned into convenient subquantities or parts, with the difference found between a bottom part and a top part.
4. Compensation principle. Sometimes it is convenient to increase or decrease the amount in one subquantity or part. This is permissible as long as the change is compensated by an increase or a decrease in amount so the value of the total quantity is not changed.

Children should be encouraged to reflect on what they are learning about computation and underlying concepts and principles. Resnick suggests that it may be helpful for students to examine and reflect upon common error patterns of other students.

> ... the process of reflecting on performance and on the relationship of procedures to underlying principles may be significantly enhanced by asking children to identify buggy procedures and explain how these procedures violate the principles of arithmetic ... [6]

In reporting his research on decimals, Swan notes that a "conflict approach," exposing inconsistencies as students rework problems in another way, may lead to deeper conceptual understanding because students discuss what a decimal is *not*.[7]

As you emphasize reflection and analysis, it may be necessary to focus on fewer examples. Even so, the instruction is likely to be more productive than having children merely practice procedures that make little sense to them. Your role as a teacher, at least in part, is to provide appropriate problems and coordinate the dicussion.

MASTERING THE BASIC FACTS

A problem encountered frequently when helping a child learn to compute is the child's inability to recall the basic facts of arithmetic without resorting to an inefficient procedure. The *basic facts* of arithmetic are the simple, closed number sentences we use when we compute. These number sentences involve two one-digit addends if they are basic addition or subtraction facts, or two one-digit factors if they are basic multiplication or division facts. They are sometimes called the "primary facts." Examples of the basic facts of arithmetic include the following:

$$6 + 7 = 13 \qquad 12 - 8 = 4 \qquad 3 \times 5 = 15 \qquad 27 \div 9 = 3$$

Children explore the basic facts in the context of learning what the operations of arithmetic mean. For example, addition can be thought of as the operation that will tell us the total number in a set if we know the amount in each of two disjoint subsets. Multiplication can also be conceived as an operation that tells us the total amount whenever we know two numbers. For multiplication, however, the two numbers are the number of equivalent disjoint subsets, and the number in each subset.

Initially, individual basic facts can be approached by younger children as problems to solve. The facts are typically presented as open number sentences such as $5 \times 4 = \Box$. Basic multiplication facts were included in a second-grade study of cooperative group problem solving by Yackel, Cobb, and Wood. Their report illustrates how children build on each other's ideas in the problem solving process.[8]

Before mastery activities are provided, children need to be taught efficient thinking strategies and more mature ways to determine the missing number. In a very useful chapter of an NCTM yearbook, Rathmell describes strategies at several different levels of maturity.[9] For addition he lists: counting on, one more or one less than a known fact, and compensation to make a known fact. For multiplication he includes: skip counting, repeated addition, one more set, twice as much as a known fact, facts of five, and patterns. It is also true that teaching distribution of multiplication over addition (multiplying "in parts") can help children proceed independently when solving untaught or forgotten multiplication combinations. By teaching such thinking strategies we help a child see relationships. Thereby, the child is able to use what he already knows as he tries to recall the missing number.

Strategies for teaching and learning addition and subtraction facts are summarized in an article by Thornton and Smith. They observed and described a successful basic facts instructional program.[10] Strategies were associated with specific sets of basic facts, and among the strategies used were count ons, count backs, doubles, near doubles, zero facts, pattern 9s, and turnarounds (commutatives). The authors observed that

> ... teachers delayed subtraction until most children had mastered "easy" addition facts. This tactic eventually resulted in mastery of a greater number of addition facts and in the children's greater (and earlier) use of addition to solve subtraction problems.[11]

Of course, when addition and subtraction are properly understood known addition facts can be used to supply a missing addend, whether in a subtraction fact or within multi-digit computation.

Mastery of the basic facts of arithmetic is the ability to supply missing sums, addends, products, and factors for basic facts promptly and without hesitation. The child who has mastered the basic fact $6 + 8 = 14$ will, when presented with "$6 + 8 = ?$," simply recall 14 without counting or figuring it out.

When a child is learning how to find the product of *any* two whole number factors (such as 36×457) by using one of the paper-and-pencil multiplication procedures, lack of mastery of the basic multiplication facts results in the child using time-consuming and distracting ways of finding needed basic products. As a result, the child's attention is drawn from the larger task of thinking through the computational procedure. Uncertain about where she is in the process, she proceeds at random, having lost her way. If a child is to use arithmetic to solve quantitive problems, it is important that the basic facts of arithmetic be mastered.

To say that mastery is important should not imply that it is necessarily easy. Consider the child who is in need of corrective instruction. In all probability this child has already met considerable failure in learning to compute and regards the whole matter with notable anxiety. Biggs observed that "in arithmetic and mathematics, the inhibition produced by anxiety appears to swamp any motivating effect, particularly where the children concerned are already anxious; or to put it another way, anxiety appears to be more easily aroused in learning mathematics than it is in other subjects."[12]

If the child has not mastered the basic facts of arithmetic, the child probably persists in using counting or elaborate procedures for finding the needed numbers. The child may seem to understand the operations on whole numbers in terms of joining disjoint sets, repeated additions, etc., yet continues to require the security of counting or other time-consuming procedures. Baroody recommends that diagnosis *begin* with a careful look at such informal procedures.[13]

A child using counting probably does not feel confident simply to recall the number. If the child practices computing, he will only reinforce his use of less-than-adequate procedures. The danger of extensive practice has been noted when the processes being practiced are developmental ones rather than the efficient practices desired. What is needed is practice *recalling* the number.

How then can you provide instructional activities that create an environment in which a child feels secure enough to try recalling missing numbers? Games provide the safest environment for simple recall, because when playing games, someone has to lose. Whereas the teacher always seems to want "the right answer," in a game it is acceptable to lose at least part of the time. The competition in a game encourages a child to use the least time-consuming procedure and to try simply recalling needed numbers. Further, games often make possible greater attending behavior by a child because of the materials involved in the game itself. For example, a child who rejects a paper-and-pencil problem such as "6 + 5 = ?" because it is a reminder of failure and unpleasantness may attend with interest when the same problem is presented with numerals painted on brightly colored cubes which can be moved about.

Obviously, what is intended is *not* the kind of arithmetic game modeled after an old-fashioned spelling bee designed to eliminate the less able children. Nor is it a game designed to put a child under pressure in front of a large group of peers. The best games will be games involving only a few children, preferably children of rather comparable ability. (If children are not of comparable ability, some may need to be given handicap points.) In such games a child can feel secure enough to try simple recall. Choose games that provide immediate or early verification. The child should learn as soon as possible if he did indeed recall correctly. Many commercial games are available, but games can be made up using materials already in the classroom or with simple materials.[14] Children are quite capable of making up games and altering the rules of games to suit their own fancy when they are encouraged to do so. For example, a homemade game using mathematical balances would provide immediate verification for each child's response (Figure 2.1).

When working privately with a child who is very insecure and insists on continuing to use elaborate procedures for "figuring out" basic facts, it may help to change the ground rules and dismiss "getting the correct answers." Try letting the child say the first number thought of after hearing or seeing the problem. Chances are the answer will be correct many times and you can show considerable surprise that the child "pulled the correct answer right out of her head." In such a setting, some children have been helped to practice recalling where they did not feel able to recall before.

FIGURE 2.1 Mathematical balance game

Games are also useful for skill retention once the basic facts are mastered. Research using the games MULTIG and DIVTIG suggests that relatively infrequent use of games can maintain skill with basic facts.[15]

Calculators will not only supply answers, but they can be used to help children learn their basic facts. For example, the constant function on a calculator can be used to help children generate products. For the products of 6 and numbers 2 through 9, children press 6 $+$ 6 $=$, $=$, $=$, etc. Make sure children do not lose count; as they press the keys, have them say: 2 sixes is 12, 3 sixes is 18, etc.

Calculators can also be used by individuals as they practice recalling the basic facts. For example, have a child say "six times seven" as he presses 6×7 . Then, have him put his hand behind his back and say "equals 42" *before* he presses $=$. The child receives immediate confirmation that he was correct. If he was incorrect, he should repeat the complete procedure immediately.

Do not infer from what has been said about mastery as a goal and the use of games and calculators that children should merely be given number sentences with missing numbers and expected to "memorize their facts." Games and calculator activities provide reinforcement which must *follow* developmental instruction. Children who have not mastered the basic facts may not have actually received the instruction that is needed.

STRESSING ESTIMATION

A proper emphasis on estimation during instruction will eliminate much of the need for future corrective instruction, so try to establish an estimation mind-set among your students.

One of the recommendations of the Lankford study, cited in Chapter 1, was that teachers "give more attention to teaching pupils to check the reasonableness of answers."[16] Those who caution that the widespread availability of calculators will not eliminate the need for computational skills also stress the need for skill in estimation. "The calculator is designed to do only the key-puncher's bidding. Nor will the calculator tell whether or not an answer is reasonable. Estimation to judge the reasonableness of an answer will still require computational skill."[17]

If children are to gain skill with estimation, they must acquire a robust "number sense" in which numeration concepts are understood and applied, the basic facts of arithmetic are easily used, as are compensation principles and other relational understandings. Estimation must be a major part of the content of mathematics instruction and not merely a procedure to verify the reasonableness of answers.

Many problem situations require an approximation or estimate rather than an exact computation. Children need to be able to distinguish among situations which call for estimation, mental computation, a paper-and-pencil procedure, or for a calculator; and students need to be prepared to take the appropriate action.

Estimation can be emphasized by allowing children time to guess, test their guesses, and revise their guesses as needed. Flexible thinking is required, and children can be encouraged to develop their own ways of deciding when an answer is reasonable. As children develop a more robust number sense, they naturally gain in their ability to estimate. In her summary of research on estimation and number sense, Sowder concludes:

> We should take advantage of the natural development of estimation in adolescents and encourage them to use estimation in many contexts in order to counteract the 'rote' tendency they acquire on other tasks. Instructional programs on estimation should take advantage of spontaneous mathematical intuitions and their development.[18]

It may also be desirable to introduce students to more standard procedures. Current elementary school texts present specific procedures for estimating answers even before computational procedures are taught. Other commercial aids for teaching estimation skills have become increasingly available. Some of these are valuable resources, filled with useful ideas.[19]

It is sometimes helpful to think of estimation as a complex of skills, any one of which may require instruction. Included among such skills for whole numbers are:

1. Adding a little bit more than one number to a little bit more than another; a little bit less than one number to a little bit less than another; and, in general, adding, subtracting, etc., with a little bit more than or a little bit less than.
2. Rounding a whole number to the nearest ten, hundred, etc.
3. Multiplying by ten and powers of ten in one step.

4. Multiplying two numbers each of which is a multiple of a power of ten (e.g., 20 × 300). This should be done as one step, without the use of a written algorithm.

With fractions, estimation often involves identifying a particular fraction as close to zero, one-half, or one.

Children's attitudes toward estimation are also important. Typically, students believe "there is only one right answer;" but when estimating, there are only reasonable answers—and some answers are more reasonable than others.

To practice estimating, children can be presented with a problem and several answers. They can then choose the answer that is most reasonable. If appropriate, their choice can be verified by computation. The practice of recording an estimated answer before computation should be encouraged. In general, children become more and more able to determine when an answer is reasonable as they gain the habit of asking if the answer makes sense, and as they progress from guessing to educated guessing to more specific estimating procedures. Children who have the habit of considering the reasonableness of their answers are not as prone to adopt incorrect computational procedures.

USING MODELS AND MANIPULATIVES

When an error in computation is observed, it is not usually sufficient to tell the child how to do it correctly. Help with the mechanics of notation is sometimes sufficient (for example, a digit can be written larger or smaller, or in a different location so as to be less confusing) but most of the time an error pattern suggests an incorrect learning that is unchallenged by the child. Often, manipulatives can be used to help the child understand the steps in the procedure, and why they make sense. Many students appear to profit from a greater number of experiences associated with the right hemisphere of the brain, and modeling with manipulatives is associated with right-hemisphere functioning.

You set the tone for your mathematics class. If you believe that getting correct answers is all that is really important, your students will believe that it is all right to push digits around even if what they are doing makes no sense to them; and when you use manipulatives your students are likely to push them around too, without relating them to a meaningful recorded procedure. From his analysis of the use of concrete materials in elementary mathematics, Thompson stresses that students "must first be committed to making sense of their activities and be committed to expressing their sense in meaningful ways."[20] Try to create the expectation throughout all of your mathematics instruction that we want to make sense of the procedures we use, and whatever we write down.

In general, well-chosen manipulatives can create an environment of what Kerslake calls "natural activity" upon which a child can build an understanding of concepts and operations. Because whole numbers are a normal part of a child's environment, young children typically create their own informal procedures while working with them. But because fractions are not a normal part of a child's environment (with few exceptions) Kerslake did not observe children creating their own informal methods; instead, children relied on rote memory of what were often half-remembered rules.[21] Materials which you can use to create more natural environments include Chip Trading Activities for whole numbers and Fraction Bars for fractions. Extensive modeling with materials like these during the early phases of instruction usually helps children develop meanings they can apply flexibly.

When you choose manipulatives to model concepts or procedures, make sure that whatever you select or design yourself is accurate mathematically. Fraction representations, for example, are sometimes inaccurate. Although it is good for children to construct many of the models used, make sure fractional parts are equal in area.

Keep in mind the need for varied models. It has already been noted that children look for commonalities among their contacts with an idea or an algorithm, and as they come to understand an idea or a procedure, they pull out or abstract the common characteristics among their experiences with that concept or procedure. Therefore, children need experiences in which all perceptual stimuli are varied except those which are essential to the mathematical idea or algorithm. A cardboard place-value chart may be of great value, but it should not be the only manipulative you use for numeration activities. Other concrete models should also be used, possibly devices made with juice cans or wooden boxes. Help your students generalize the desired concept or procedure rather than ideas or processes specific to a particular model. For example, students need to learn that "ten ones is the same amount as one ten," rather than "take ten yellows to the bank then put one blue in the next place." You also need to vary the examples children encounter when they work at the symbolic level. In the subtraction number sentence $42 - 17 = 25$, a child may conclude that the five units in the answer is simply the result of finding the difference between the two and the seven. Examples should be varied so that irrelevant characteristics are not observed as common attributes.

For students who adopt error patterns in computation, it is often wise to redevelop computational procedures as careful step-by-step records of observations while using manipulatives. Hopefully, this will help the child who has been pushing symbols around in a rote manner. It will help him make sense of his record—the algorithm itself. For most whole number algorithms, it is possible to manipulate models and record the actions step-by-step so that the desired computational procedure is the resulting record. Sometimes the manipulatives can even be arranged in relation to one another just as digits are placed in relation to one another when computing on paper. This is true for some of the activities suggested in Chapter 5. Sticks (singles and bundles), base-ten blocks, and place-value charts are often used. For algorithms with fractions, especially unlike fractions, it is not always possible to demonstrate each step in the computational procedure with manipulatives. Often, the procedure must be developed by reasoning with mathematical ideas; then manipulatives are used to verify the result previously obtained.

Finally, whenever you have students use manipulatives to model concepts and procedures, involve your students in experiences which "go both ways."

That is, have students manipulate models and record what they observe with symbols, but also have students begin with symbols and interpret the symbols by modeling the concept or procedure. For example, to find how many sets of 3 cubes can be made from a set of 14, a student may make sets of 3 cubes and count them (observing that 2 cubes remain). The student then records what he has observed as

$$14 \div 3 = 4r2, \text{ or as} \quad \frac{4r2.}{3\overline{)14}}$$

Also, to "go the other way" this student can be given a number sentence like $17 \div 3 = 5r2$ and asked to show what it means with cubes; that is, model it.

TEACHING CHILDREN TO COMPUTE

As adults we compute in different ways. The particular method that is appropriate depends upon the situation. We may compute mentally, we sometimes use a paper-and-pencil procedure, and at other times we use calculators. Children also need to be able to compute in each of these ways. As students learn these different methods, they need practice in choosing the appropriate method for particular situations.[22]

It is usually wise to introduce a paper-and-pencil algorithm by presenting children with interesting verbal problems and letting them use what they already know to work out solutions. Or, simply present an example and ask the students how they think it should be solved. Be careful not to use direct instruction too soon.

> ... the exclusive use of direct instruction by teachers in the early grades to
> teach mathematical algorithms and procedures may result in children learning
> by rote the mathematics algorithms and procedures, but not acquiring a true
> mathematical understanding. Children may not see the connection between
> their informal knowledge and the formal mathematics they are taught in school.
> This may result in formal mathematics being reduced to meaningless symbol
> manipulation.[23]

If you let students use their own informal techniques initially, you will be surprised to learn that some children know more than you thought. Other children will be very creative in using what they know. By beginning in this way, you help children relate the algorithm you are teaching to the informal knowledge of arithmetic they already possess.

Before you teach children to compute with paper, make sure your students are well grounded in principles that permit them to represent quantities in multi-digit decimal notation and to also make appropriate exchanges. You can use Chip Trading Activities (or teacher-made variations) to help children develop these abilities. Then use instruction in computation as an opportunity to further develop each child's "number sense."

In the normal developmental sequence for teaching an algorithm for whole numbers, children usually begin by solving problems with the aid of manipulatives. Eventually, a step-by-step record of manipulations and thinking is written with numerals. If you keep in mind the algorithm you are working toward as you guide the recording, the written record will become the algorithm itself. After children are comfortable with this procedure, they visualize or refer to the manipulatives (but they do not actually handle them) as they write.

In some cases, in order to derive the standard algorithm you will need to gradually shorten the written record. You may want to say that "mathematicians like to write fewer symbols whenever they can."

Admittedly there are algorithms that cannot be developed as a record of observations. These computational procedures can sometimes be introduced as a shortcut. For example, the standard procedure for dividing fractions can be developed by reasoning through a rather elaborate but meaningful procedure involving complex fractions, application of the multiplicative identity, and the like, and then observing a pattern. The obvious implication of the pattern is that most of the steps can be eliminated by merely inverting the divisor and multiplying.

It is the experience of the author that when manipulatives are used to teach an algorithm, the critical step is progressing from the manipulatives to written symbols. This is why a *step-by-step* record is helpful. The record or algorithm must make sense if the child is going to do more than push symbols around on paper. Even textbooks typically explain the rationale for renaming when adding and subtracting (sometimes called carrying and borrowing) and they present helpful drawings. But Resnick cautions " . . . instructional attention passes quickly to efficient calculation, thereby probably encouraging automation of calculation rules that are not well linked to . . . principles."[24] Resnick further cautions that such understandings do not always prevent error patterns. We must be careful not to let children make procedures automatic until they are correct, or we are likely to observe error patterns.

> . . . simply explaining and demonstrating the principles of place value arithmetic to children would not have much of an effect on their calculation performance. Even improving children's understanding to the point where they could construct explanations themselves could not be counted on to eliminate buggy calculation rules once children had adopted a more or less automatic procedure.[25]

During good developmental instruction, *continuing* diagnosis is very important. Say to children repeatedly, "Tell me something about this," and thereby help them develop the ability to communicate mathematical ideas as well as give you diagnostic information. Be careful not to overtest at-risk students, but keep your eyes open.

One fruit of continuing diagnosis is feedback for your students. Research indicates that students need corrective feedback that not only tells students which examples are correct or wrong, but also provides assistance in obtaining correct answers. Corrective feedback can take many forms. It can be oral, along with comments to the child which express confidence in his ability to learn, or it can be written on the student's paper. Bloom and Bourdon studied the written feedback techniques used by classroom teachers, and found that frequently teacher feedback did not incorporate *corrective* feedback. Often, papers were merely scored and students were asked to rework the examples. Some form of corrective assistance needs to be written on the paper as well.[26]

It is hoped that paper-and-pencil computation will not dominate the mathematics curriculum in future years. Instead, it is hoped that students will have time to learn all the mathematics they need. As stated in the 1989 *Curriculum and Evaluation Standards for School Mathematics* of the National Council of Teachers of Mathematics, we need to have "reasonable expectations for proficiency with paper-and-pencil computation."[27]

Many children will make mistakes while learning to compute. Even so, mistakes can be an important, positive part of the initial learning process. But teachers respond differently to errors in different cultures.

> We have been struck by the different reactions of Asian and American teachers to children's errors. For Americans, errors tend to be interpreted as an indication of failure in learning the lesson. For Chinese and Japanese, they are an index of what still needs to be learned. These divergent interpretations result in very different reactions to the display of errors—embarrassment on the part of American children, calm acceptance by Asian children. They also result in differences in the manner in which teachers utilize errors as effective means of instruction.[28]

Your attitude toward errors is important. View them as opportunities for learning.

Continually emphasize estimating answers and this will help children discover many of their errors. When a child who is learning to compute produces an unreasonable answer, the child faces a personal problem-solving situation which can lead to creative thinking and to a greater understanding of the algorithm itself. Whenever children discover that their ideas do not produce correct or reasonable answers, they should be motivated to resolve the dilemma rather than be penalized for the mistake. Use errors as springboards for the learning of mathematics.

Be alert to any perceptual difficulties a child may have. In order to respond to instruction, children need to be able to observe and also envision the physical properties of digits: vertical vs. horizontal elongation, straightness vs. curvature, and degree of closure. Because instruction emphasizes understanding, children also need to be able to perceive attributes of multi-digit numerals—properties such as position of a digit to the left or right of another digit. Also, poor spatial ability may affect a child's capacity to respond to instruction emphasizing place value concepts.

Other children may find it difficult to respond to instruction because of language patterns. The syntax of English language expressions is often different from the structure of mathematical statements, and we complicate the situation by using different but equivalent language expressions for the same situation. (For example, *twelve minus four* and *four from twelve* express the same mathematical concept.) Researchers are studying the implications for instruction.

While conducting research with both Hispanics and Anglos, Mestre and Gerace found, for example, that we need to focus careful attention on helping students discern the differences between labels and variables.[29] In her book *Twice As Less*,[30] Orr raises questions about the effects of nonstandard English on students' learning of mathematics. She concludes that nonstandard language is often associated with erroneous concepts, and she urges teachers to replace rote memorization of procedures with an emphasis on understanding and verbalization of mathematical concepts, relationships, and processes.

After a child gains some competence with an algorithm, less stress should be placed on mathematical understanding. As Wheatley notes, "The advantage of algorithms lies in the routine established and the minimum amount of decision making or problem analysis required."[31] It is primarily while a computational procedure is being learned that understanding plays a role. After it is learned the greater concern is with the meaning of the arithmetic operation itself, so that children will know when it is appropriate to *use* the algorithm. For example, What is given in a problem situation? What is wanted? If a sum and an addend are given and the other addend is wanted, then a subtraction algorithm can be used to find that missing addend.

Thorough, developmental instruction is needed in which it is possible for each child to move through a carefully planned sequence of different types of learning activities. The amount of time needed for each type of activity will vary from child to child, and, for any one individual, the pace will likely vary from day to day. If you are to lessen the likelihood that children learn patterns of error, you will have to resist the temptation to cover the text or the curriculum guide by completing two pages a day or some similar plan. Careful attention will have to be given to ideas and skills needed by each child in order to learn the concept or algorithm under study.

Teach in a way that makes the adoption of erroneous procedures a very unlikely event!

USING CALCULATORS AND COMPUTERS

Those who expect to see calculators used increasingly for routine calculations believe students should be encouraged "to perform computation without the calculator if the problem is more efficiently completed that way."[32] If children are to be able to judge the most efficient procedure in a given situation, they will need to know how to use a calculator, how to do paper-and-pencil calculations, and how to compute mentally. They will need practice deciding which is most appropriate in different situations.

The National Council of Teachers of Mathematics, while recommending that mathematics programs "take full advantage of the power of calculators," also recognizes that "a significant portion of instruction in the early grades must be devoted to the direct acquisition of number concepts and skills without the use of calculators."[33] However, a calculator is not simply an alternative to paper-and-pencil procedures; it can help children *learn* those procedures.

A calculator can be used to reinforce underlying concepts and procedures—especially numeration concepts. For example, students can practice naming what some have called the "product value of a digit" (face value × place value). Consider these instructions:

> Enter "4,653" into your calculator.
> Now, use addition or subtraction to change the six to a zero. You should have "4,053."[34]

Basic multiplication products can be generated by using the repeat function of calculators.

6 × 7 = ? Think of 6 × 7 as 6 sevens.

Key ⑦ ⊞ ⑦ ⊟ ⊟ ⊟ ⊟ ⊟

[counting 2, 3, 4, 5, 6,]

A calculator can be used to focus on one step in an algorithm: for example, placement of partial products in multiplication. Calculators can also be used for estimating quotients in division.

> Provide several examples similar to 83,562 ÷ 36 or 17,841 ÷ 892, then have students select one example.
> Each person estimates the answer, and writes it; then the exact answer is determined with a calculator.
> Players score one point if the correct number of digits is in their estimate; but they score two points if they have the correct number of digits and the first digit is also correct.[35]

Calculators can be used to provide immediate feedback when children practice the basic facts of arithmetic, and they can be used to check answers when children compute with paper. Yes, a calculator has many uses—but its limitations must also be demonstrated. For example, it takes more time to multiply by a power of ten on a calculator than to perform the multiplication mentally.

Quality computer software for instruction in mathematics is increasingly available. As far as computation is concerned, computer programs should be carefully selected. Make sure the software is the best medium available for the particular phase of instruction, because whenever a paper skill is being taught students will need to use paper during some of the instructional process.

Whenever children work at a computer, talk with them from time to time. Have them explain what they are doing and why specific choices are made.

DEVELOPING FLOWCHARTS

Whenever you teach computation procedures, whether during regular developmental instruction or as you intervene with students experiencing difficulty, student-generated flowcharts can enhance instruction.

To teach simple flowcharting procedures, make a chart available for reference which shows the basic shapes and how they are used. Figure 2.2 is an example of such a chart. Focus on procedures for simple mathematical tasks as you develop one or more charts with your students. Figure 2.3 is an example of a flowchart developed to show how to choose the greatest of several whole numbers. Flowcharts that are developed can be tested by having children in a different group (or classroom) follow the chart step-by-step.

After students have learned how to make a flowchart, have pairs of students, and eventually individuals, create flowcharts for paper-and-pencil computation procedures. Figure 2.4 is a flowchart for adding fractions prepared by a fourth-grade boy. Be sure that each child tests his or her flowchart by having at least one other student work through the chart step-by-step.

Flowcharts will provide diagnostic feedback for you and for each student. Of course, this is a part of the instructional process. As flowcharts are tested, compared, and discussed with other students and with you, the teacher, the charts should be remade until they are correct and complete.

When the end of a school year approaches, some teachers have their students summarize computational procedures they have learned during the year by making flowcharts they can take with them to their new classroom or school.

USING ALTERNATIVE ALGORITHMS

If you are really willing to accept the idea that there are many legitimate ways to subtract, divide, and so on, you can choose to introduce an algorithm that is fresh and new to the student experiencing difficulty. By so doing, you may circumvent the mind-set of failure which beleaguers the child. But if you are convinced that there is really only one "right" way to subtract or divide, you will

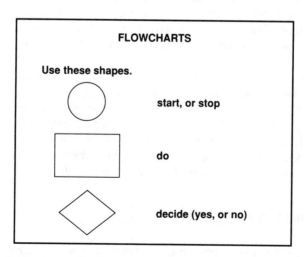

FIGURE 2.2 Basic shapes for flowcharts

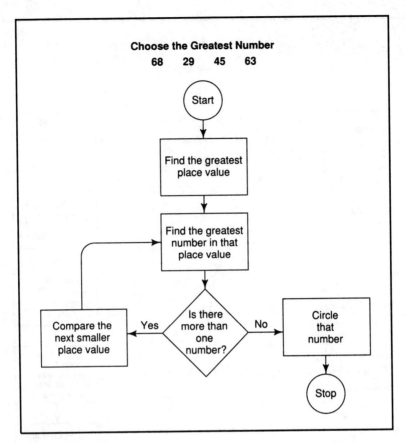

FIGURE 2.3 An example of a flowchart

have no heart for introducing alternative procedures. When the immediate need of the child is for successful experience in mathematics, the use of alternative algorithms holds promise, but you should first face yourself squarely and make sure you can heartily endorse such procedures as having value in themselves apart from any "right" way of computing. If you believe you are leading the child to an inferior technique, the child will sense your feeling.

An important question arises when you decide to show a child an alternate computational procedure, because many of the algorithms which have been employed through the centuries have been used much as you would use a machine—without the user knowing why the procedure produces the correct result. Indeed, some of the algorithms of historical interest are quite difficult to explain to children. Although many of the alternatives available can be shown to make sense, in your work with an unsuccessful child some of the alternatives may be learned only as a machine to use when a result is needed. Do you present only algorithms which are likely to make sense to the child, or do you sometimes try to teach a child how to use a machine?

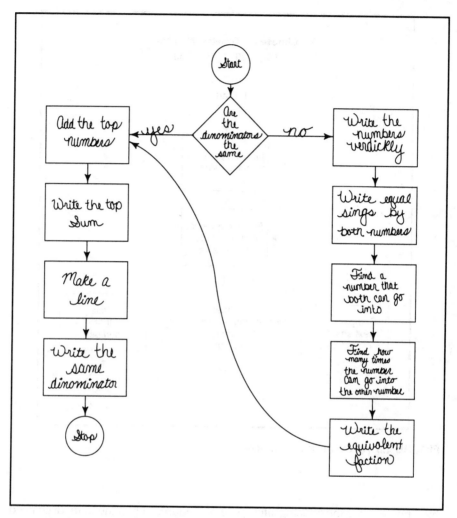

FIGURE 2.4 A fourth grader's flowchart for adding fractions

I am convinced of the need to teach computational procedures that make sense to a child. In view of the evidence reported in our research literature, any other conviction would hardly be warranted. Meaningful instruction is needed not just for the child who learns easily, but also for those children who experience difficulty when learning computation procedures. All children need to make connections; it facilitates recall of information. Also, when students understand relationships they are more likely to apply what they learn.

At the same time, when you work with a child who has known much failure, that child will not readily attend to instruction similar to past unhappy experiences. In order to secure the interest and attention of such a child, a very different experience is needed. It may be that the *most* important need for the child is to get a *correct* answer, or to experience success. In view of these considerations, there are times when a computational procedure *that is fresh and new to the child* can be taught as a machine to use whenever a result is needed. If the child does learn and remember the procedure, the needed success experience has been provided. If the child does not, then the procedure can be set aside much as you would set aside anything else you try and do not like. Interestingly, alternative algorithms have always been explored unhesitatingly as enrichment or recreational activity.

What are some of the algorithms that can be used as alternatives when working with a child who is experiencing difficulty? A few examples are described here, and others can be found in the literature of mathematics education and recreation.[36]

Some of the algorithms that follow are low-stress algorithms. When working with children who are having difficulty learning to compute you may find that low-stress algorithms are learned more readily than alternatives. Like other algorithms, they can often be taught as sensible records of manipulations with sticks, blocks, or number rods. The advantage of low-stress algorithms is that they separate fact recall from renaming, and thereby place fewer memory demands on the child who is computing.

Addition of Whole Numbers: Hutchings's Low-Stress Method[37]

As illustrated in examples A and B, sums for basic facts are written with small digits to the left and to the right instead of in the usual manner.

When a column is added, the 1 ten is ignored and the one's digit is added to the next number. In example C, $5 + 9 = 14$, $4 + 8 = 12$, $2 + 4 = 6$, and $6 + 7 = 13$. The remaining 3 ones are recorded in the answer. The 3 tens are counted and also recorded. Multi-column addition proceeds similarly, except the number of tens counted is recorded at the top of the next column.

Many of the children for whom regrouping in addition is difficult find this alternative algorithm rather easy to learn. It often brings success quickly, even with large examples. Other children, especially those with perceptual problems, may be confused by the abundance of crutches. This is especially true when it is not possible to write examples with large digits, which is the case with most published achievement tests.

Addition of Whole Numbers: Left-Handed Addition[38]

Before 1900, left-handed addition was often used to check sums obtained from right-handed addition. Today this form of "front end" addition may be used to help children focus on reasonable solutions—to help them learn to estimate.

Problem:

$$+\ \begin{matrix} 6 & 9 & 8 \\ 5 & 7 & 9 \end{matrix}$$

$$\begin{matrix} 1 & 1 \\ & 1 & 6 \\ & & 1 & 7 \end{matrix}$$

Sum: $1\ 2\ 7\ 7$

Partial sums are written left-to-right as place-value columns are aligned carefully. Then addition proceeds again, left-to-right, and the sum is written with standard notation.

Subtraction of Whole Numbers:
The Equal Additions Method

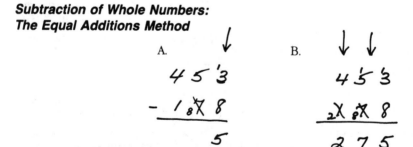

A.
$$\begin{matrix} 4 & 5 & ^1 3 \\ -\ 1 & _8 \cancel{7} & 8 \end{matrix}$$
$$ 5$$

B.
$$\begin{matrix} 4 & \cancel{5}^1 & 3 \\ _2\cancel{1} & _6\cancel{7} & 8 \end{matrix}$$
$$2\ 7\ 5$$

The principle of compensation is applied: when equal quantities are added to both the minuend and the subtrahend, the difference remains the same. In this computation, ten is added to the sum, i.e., to the three in the ones place. To compensate for this addition, ten is also added to the known addend; the seven in the tens place is replaced with an eight. Similarly, 1 hundred is added to the sum, i.e., the five in the tens place becomes a fifteen. To compensate, one hundred is added to the known addend; the one in the hundreds place is replaced with a two.

Multiplication of Whole Numbers: The Lattice Method

Problem:
$$\begin{array}{r} 627 \\ \times 354 \end{array}$$

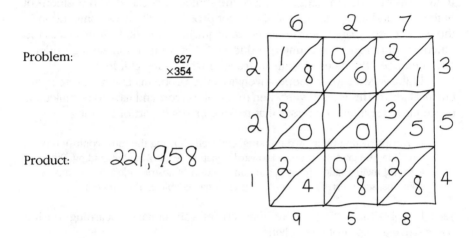

Product: 221,958

The two factors are written above and to the right of a grid. The products of basic multiplication facts are recorded within the grid, and addition proceeds from lower right to upper left within the diagonal lines. The final product is read at the left and at the bottom of the grid.

Subtraction of Rational Numbers: The Equal Additions Method[39]

Problem:

$$-\begin{array}{r} 7\frac{1}{4} \\ 3\frac{3}{4} \\ \hline 3\frac{1}{2} \end{array}$$

add one $\left(\frac{4}{4}\right) \longrightarrow 7\frac{5}{4}$

add one $(1) \longrightarrow \div 4\frac{3}{4}$

Difference: $3\frac{2}{4}$

The principle of compensation is applied. One is added to both the minuend and the subtrahend in order to subtract easily.

USING COOPERATIVE GROUPS

Children profit from involvement with other children while working cooperatively on a problem or task. This is true for all children, including students who experience difficulty learning computation procedures. On-task behavior often improves among students who work cooperatively in groups.

A focus on cooperation reflects values that are different than values reflected by a focus on competition among students. In *The Competitive Ethos and Democratic Education,* Nicholls asserts that in a society like ours in which some individuals must acknowledge they lack the abilities associated with positions of high status and accept lesser roles, it is not productive to stress competition in the early school years when students are acquiring basic skills. It produces too much alienation and it is counterproductive.[40] Yet competition among students has often been the focus of skills instruction with young children.

Children who work cooperatively to solve problems communicate mathematical ideas. This is true even when they make errors and have to rethink their procedures. Labinowicz stresses that making errors is part of learning.

> In reconstructing their own thinking, children showing the most confusion often show the greatest progress toward a mature level of understanding. Since learning mathematics involves the construction of abstract ideas and complex relations, some errors and confusion are unavoidable in the process.[41]

Yackel, Cobb, and Wood note that errors can enhance learning during cooperative-group problem solving.

> If a child's explanations are based on the partner's erroneous answer, the explaining child must extend his or her own conceptualization of the problem to try to infer how the partner might have been trying to make sense of the problem.[42]

But if you use cooperative groups as an instructional strategy, how should you constitute the working groups of children?

When you include within a cooperative group a student who is experiencing difficulty learning computational procedures, put the child in a problem-solving situation with more capable peers—or even with an adult. At the same time, be sure that participants believe they are responsible for their own learning and also for the learning of others in the group. They need to be individually accountable (each should be able to demonstrate mastery of the content) but they also need to be accountable for others in the group.[43]

What kinds of tasks or problems should you design? When skills are involved, structure group work so that children explain procedures to one another. Participants in the group need to make sure each child in the group understands and can do the required procedures.[44] Some tasks or problems may involve manipulatives, but remember that children do not learn by just using manipulatives; they learn by *thinking about what they are doing* when they use manipulatives. Be sure they reflect on what they are doing, and explain it to one another. It has been said that . . .

> We should never tell a child what that child might be able to tell us. Similarly, we should never tell a child what some other child might be able to tell the child for you.

Sometimes your assignment for a group can be for them to create a task or problem which requires a particular computation. As you observe children working on such an assignment, focus on their thinking—not on their answers alone. At other times when you give students a problem to solve, have them demonstrate two or more procedures for solving it.

Sample problems for cooperative groups are described in Appendix E. The particular tasks focus on computational procedures.

GUIDING INSTRUCTION

Teachers have found the following guidelines to be helpful. They provide a summary of principles to keep in mind when instructing children who are having difficulty learning to compute.

Focus on the Child

1. *Personalize instruction.* Even when children meet in groups for instruction, individuals must be assessed and corrective programs must be planned for *individuals.* Some individual tutoring may be required.

2. *Believe the child is capable of learning.* A student who has met repeated failure needs to believe that she is a valued person and is capable of eventually acquiring the needed skill. You must believe this also if you are to help the child. Ginsburg notes that, "In some cases, helping children to improve their schoolwork may do more for their emotional health than well-meaning attempts to analyze and treat their emotional disturbances directly."[45] Remember that if children are to learn, they need an environment in which they are perceived as capable of learning.

3. *Make sure the child has the goals of instruction clearly in mind.* Take care to ensure that the behavior that is needed on the child's part is known; the child needs to know the direction instruction is heading. He needs to know where it will head eventually ("I'll be able to subtract and get the right answers"), but also, where it is headed immediately ("I'll soon be able to rename a number many different ways").

4. *Encourage self-appraisal by the child.* From the beginning, involve her in the evaluation process. Let the child help set the goals of instruction.

5. *Assure consistent expectations.* Make sure that you and the child's parents have the same expectations in regard to what the child will accomplish. People in the United States tend to assume that difficulties with learning result from a lack of ability; in many other countries, they are more likely to assume that

difficulties are a result of insufficient effort. Be sure that you and the child's parents are together in regard to such expectations.

6. *Provide the child with a means to observe any progress.* Charts and graphs kept by the child often serve this function.

Teach Concepts and Skills

7. *Build on underlying concepts and procedures the child knows.* Corrective instruction should build on a child's strengths; it should consider what the child is ready to learn. Typically, children need a thorough understanding of subordinate mathematical concepts before they can be expected to integrate them into more complex ideas.

8. *Emphasize ideas that help the child organize what he learns.* Children often assume the concept or procedure they are learning applies only to the specific task they are involved in at the time. Connect new learnings with what a child already knows. When so organized, new learnings can be more easily retrieved from a child's memory as the need arises; also, they can be more readily applied in new contexts. Stress ideas such as multiple names for a number, commutativity, identity elements, and inverse relations.

9. *Stress the ability to estimate.* A child who makes errors in computation will become more accurate with the ability to determine the reasonableness of answers.

Provide Instruction

11. *Base instruction on your diagnosis.* Take into account the patterns you observed while collecting data. What strengths can you build upon?

12. *Choose instructional procedures that differ from the way the child was previously taught.* The old procedures are often associated with fear and failure by the child; something new is needed.

13. *Use a great variety of instructional procedures and activities.* Variety is necessary for adequate concept development. A child forms an idea or concept from many experiences embodying that idea; the child perceives the concept as that which is common to all of the varied experiences.

14. *Connect content to experiences out of school.* A child who can tie what she is learning to experiences out of school is likely to be motivated to learn and able to apply what she does learn.

15. *Encourage the child to "think out loud" while working through a problem situation.* Have the child show how and explain why certain materials and procedures are being used. Speaking out loud often enables a child to focus more completely on the task at hand.[46]

16. *Ask leading questions that encourage reflection.* Allow sufficient time for the child to reflect.

17. *Let the child state his understanding of a concept in his own language.* Do not always require the terminology of textbooks. It may be appropriate to say, for example, "Mathematicians have a special name for that idea, but I rather like your name for it!"

18. *Sequence instruction in smaller amounts of content for children having difficulty learning.* A large task may overwhelm such a child. However, when instruction is based upon a carefully determined sequence that leads to the larger task, the child can focus on more immediately attainable goals. The child can also be helped to see that the immediate goals lead along a path going in the desired direction.

19. *Move toward symbols gradually.* Move from manipulatives to two-dimentional representations and visualizations to the use of symbols. Carefully raise the level of the child's thinking.

20. *Emphasize careful penmanship and proper alignment of digits.* A child must be able to read the work and tell the value assigned to each place where a digit is written. Columns can also be labeled if appropriate.

Use Concrete Materials

21. *Let the child choose from materials available.* Whenever possible, the child should be permitted to select a game or activity from materials that are available and lead toward the goals of instruction. Identify activities for which the child has needed prerequisite skills and which lead to the goals of instruction; then let the child have some choice in deciding what she will do.

22. *Encourage a child to use aids as long as they are of value.* Peer group pressure often keeps a child from using an aid even when the teacher places such aids on a table. The use of aids needs to be encouraged actively. Occasionally, a child needs to be prompted to try thinking a process through with just paper and pencil, but, by and large, children give up aids when they feel safe without them. After all, the use of aids is time-consuming.

Provide Practice

23. *Make sure the child understands the process before assigning practice.* We have known for some time that, in general, drill reinforces and makes more efficient that which a child *actually* practices.[47] In other words, if a child counts on his fingers to find a sum, drill will only tend to help him count on his fingers more efficiently. The child may find sums more quickly, but he is apt to continue any immature procedure he is using. You stand forewarned against the use of extensive practice activities at a time when they merely reinforce processes that

are developmental. Drill for mastery should come when the actual process being practiced is an efficient process. Admittedly, it is not always easy to determine what process is actually being practiced. By looking for patterns of error and by conducting data-gathering interviews in an atmosphere in which a child's failures are accepted, you can usually learn enough to decide if a child is ready for more extensive practice.

24. *Select practice activities which provide immediate confirmation.* When looking for games and drill activities to strengthen skills, choose those activities that let the child know immediately if the answer is correct. Many games, manipulative devices, programmed materials, and teacher-made devices provide such reinforcement.

25. *Spread practice time over several short periods.* Typically, a short series of examples (perhaps five to eight) is adequate to observe any error pattern. Longer series tend to reinforce erroneous procedures. If a correct procedure *is* being used, then frequent practice with a limited number of examples is more fruitful than occasional practice with a large number of examples.

Later, in Chapter 5, you will read about many specific suggestions for instruction. The lists of references following that chapter will point you to additional ideas.

ENDNOTES

1. Robert B. Ashlock, et. al., *Guiding Each Child's Learning of Mathematics* (New York: Macmillan, 1983), 482.
2. Herbert Ginsberg, *Children's Arithmetic: The Learning Process* (New York: D. Van Nostrand, 1977), 81.
3. *Ibid.,* 85–89.
4. Lauren B. Resnick, "Beyond Error Analysis: The Role of Understanding in Elementary School Arithmetic," in *Diagnostic and Prescriptive Mathematics: Issues, Ideas, and Insights, 1984,* Helen N. Cheek, ed., research monograph of the Research Council for Diagnostic and Prescriptive Mathematics (Kent, Ohio: The Council, 1984), 2.
5. *Ibid.,* 6.
6. *Ibid.,* 13.
7. Malcolm Swan, "Teaching Decimal Place Value: A Comparative Study of "Conflict" and "Positive Only" Approaches," in *Proceedings of the Seventh International Conference for the Psychology of Mathematics Education,* Rina Herschkowitz, ed. (Rehovat, Israel: Weizmann Institute of Science, 1983), 211–216.
8. Erna Yackel, Paul Cobb, and Terry Wood, "Small-Group Interactions as a Source of Learning Opportunities in Second-Grade Mathematics," *Journal of Research in Mathematics Education* 22 (November 1991): 390–408.
9. Edward C. Rathmell, "Using Thinking Strategies to Teach the Basic Facts" in Marilyn N. Suydam and Robert E. Reys, eds., *Developing Computational Skills: 1978 Yearbook* (Reston, Va.: National Council of Teachers of Mathematics, 1978), 13–38.
10. Carol A. Thornton, and Paula J. Smith, "Action Research: Strategies for Learning Subtraction Facts" *The Arithmetic Teacher* 35 (April 1988): 8–12.
11. *Ibid.,* 11.
12. John Biggs, "The Psychopathology of Arithmetic," in *New Approaches to Mathematics Teaching,* F. W. Land, ed. (London: Macmillan & Co., 1963), 59.

13. Arthur J. Baroody, "Children's Difficulties in Subtraction: Some Causes and Cures," *The Arithmetic Teacher* 32 (November 1984): 14–19.

14. For guidelines see Robert B. Ashlock and Carolynn A. Washbon. "Games: Practice Activities for the Basic Facts," in *Developing Computational Skills: 1978 Yearbook* Marilyn N. Suydam and Robert E. Reys, eds. (Reston, Va.: National Council of Teachers of Mathematics, 1978), 39–50.

15. George W. Bright, John G. Harvey and Margariete M. Wheeler, "Using Games to Maintain Multiplication Basic Facts," *Journal for Research in Mathematics Education* 11, no. 5 (November 1980): 379–85.

16. Lankford, 1972, 42.

17. Eugene P. Smith, "A Look at Mathematics Education Today," *The Arithmetic Teacher* 20 (October 1973): 505

18. Judith Sowder, "Estimation and Number Sense," in *Handbook of Research on Mathematics Teaching and Learning,* Douglas A. Grouws, ed. (New York: Maxwell Macmillan International, 1992), 377.

19. For example, see Robert E. Reys, Paul Trafton, Barbara Reys, and Jody Zawojewski, *Computational Estimation* (Palo Alto, Calif.: Dale Seymour Publications, 1987). Books are available for each of the grades 6 through 8. Lesson plans and masters for transparencies and worksheets are included.

20. Patrick W. Thompson, "Notations, Conventions, and Constraints: Contributions to Effective Uses of Concrete Materials in Elementary Mathematics," *Journal for Research in Mathematics Education* (March 1992): 146.

21. Daphne Kerslake, *Fractions: Children's Strategies and Errors* (Windsor, Berkshire: The NFER-NELSON Publishing, 1986), 87.

22. See "Essential Mathematics for the Twenty-first Century: The Position of the National Council of Supervisors of Mathematics, *The Arithmetic Teacher* 36, no. 9 (May 1989): 27–29.

23. Penelope L. Peterson, "Teaching for Higher-Order Thinking in Mathematics: The Challenge for the Next Decade," in *Perspectives on Research on Effective Mathematics Teaching,* vol 1, Douglas A. Grouws and Thomas J. Cooney, eds. (Reston, Va.: National Council of Teachers of Mathematics, Inc., 1986), 7.

24. Lauren B. Resnick, "Beyond Error Analysis: The Role of Understanding in Elementary School Arithmetic," in *Diagnostic and Prescriptive Mathematics: Issues, Ideas, and Insights,* 1984, Helen N. Cheek, ed., research monograph of the Research Council for Diagnostic and Prescriptive Mathematics (Kent, Ohio: The Council, 1984), 13.

25. *Ibid.,* 13.

26. Robert B. Bloom and Linda Bourdon, "Types and Frequencies of Teachers' Written Instructional Feedback," *Journal of Educational Research* 74 (September/October 1980), 13–15.

27. National Council of Teachers of Mathematics, *Curriculum and Evaluation Standards for School Mathematics,* (Reston, Va.: The Council, 1989), 44.

28. James W. Stiger, and Harold W. Stevenson, "How Asian Teachers Polish Each Lesson to Perfection," *American Educator* (Spring 1991): 27.

29. Jose Mestre and William Gerace, "The Interplay of Linguistic Factors in Mathematical Translation Tasks," *Focus on Learning Problems in Mathematics* 8 (Winter 1986): 59–72.

30. Eleanor W. Orr, *Twice As Less: Black English and the Performance of Black Students in Mathematics and Science* (New York: W. W. Norton, 1987).

31. Grayson H. Wheatley, "A Comparison of Two Methods of Column Addition," *Journal for Research in Mathematics Education* 7, no. 3 (May 1976): 145–54.

32. Thomas P. Carpenter et al., "Calculators in Testing Situations: Results and Implications from National Assessment," *The Arithmetic Teacher* 28, no. 5 (January 1981): 37.

33. *An Agenda for Action: Recommendations for School Mathematics of the 1980s* (Reston, Va.: National Council of Teachers of Mathematics, 1980), 8.

34. See the game "Wipe Out" in Wallace Judd, "Instructional Games with Calculators," *The Arithmetic Teacher* 23, no. 7 (November 1976): 516.

35. See Earl Ockenga, "Calculator Ideas for the Junior High Classroom," *The Arithmetic Teacher* 23, no. 7 (November 1976): 519.
36. For example, see Eleanor S. Pearson, "Summing It All Up: Pre-1900 Algorithms," *The Arithmetic Teacher* 33 (March 1986): 38–41.
37. Barton Hutchings, "Low-Stress Algorithms," *Measurement in School Mathematics,* 1976 Yearbook of the National Council of Teachers of Mathematics (Washington, D.C.: NCTM, 1976), 218–39.
38. Eleanor S. Pearson, "Summing It All Up: Pre-1900 Algorithms," *The Arithmetic Teacher* 33 (March 1986): 38–41.
39. Barbara Signer, "The Method of Equal Addition . . . It's Rational!!" *Dimensions in Mathematics* 5 (Winter 1985): 15–17.
40. John G. Nicholls, *The Competitive Ethos and Democratic Education,* (Cambridge, Mass.: Harvard University Press, 1989).
41. Ed Labinowicz, "Children's Right to Be Wrong," *The Arithmetic Teacher* 35 (December 1987): 2.
42. Erna Yackel, Paul Cobb, and Terry Wood, "Small-Group Interactions as a Source of Learning Opportunities in Second-Grade Mathematics," *Journal for Research in Mathematics Education* 22 (November 1991): 402.
43. See David W. Johnson and Roger T. Johnson, "Toward a Cooperative Effort: A Response to Slavin," *Educational Leadership* 46 (April 1989): 80–81.
44. See Erna Yackel, et al., "The Importance of Social Interaction in Children's Construction of Mathematical Knowledge," in *Teaching and Learning Mathematics in the 1990s,* yearbook, Thomas Cooney Jr., and Christian R. Hirsch, eds., National Council of Teachers of Mathematics (Reston, Va.: The Council, 1990), 12–21.
45. Herbert Ginsberg, *Children's Arithmetic: The Learning Process.* (New York: D. Van Nostrand Company, 1977), 197.
46. Brenda H. Manning, "A Self-Communication Structure for Learning Mathematics," *School Science and Mathematics* 84 (January 1984): 43–51.
47. See William A. Brownell and Charlotte B. Chazel, "The Effects of Premature Drill in Third Grade Arithmetic," *The Journal of Educational Research* 29 (September 1935): 17–28ff: also in Robert B. Ashlock and Wayne L. Herman, Jr., *Current Research in Elementary School Mathematics* (New York: Macmillan, 1970), 170–88.

SPECIFIC ERROR PATTERNS

In Part Two you have the opportunity to identify particular examples of erroneous learning. You also will be able to suggest what might be done to help individual students, then you can compare your suggestions with those of the author.

Specifically, in Chapter Three you will examine student papers to determine what error patterns have been learned. (Some of the examples have correct answers, though the child's thinking was incorrect.) In Chapter Four you will learn how each student computed and why he or she may have adopted the particular erroneous procedure. Then in Chapter Five you will read about ways to help each student.

You can read Part Two in either of two ways. You can read it one chapter at a time in sequence. Or, you can read about one error pattern at a time, moving through all three chapters for a given error pattern, as directed in the text.

Identifying Error Patterns in Computation

Instruction that responds to incorrect procedures should be based on sufficient data to divulge *patterns* of incorrect and immature computations. As a teacher you need to be alert to error patterns.

On the following pages you will find examples of the written work of boys and girls who are having difficulty with some phase of computation. With these simulated children's papers you have the opportunity to develop your own skill in identifying error patterns. However, these are more than simulated papers. The papers contain the error patterns of real boys and girls observed by teachers in ordinary school settings. The children are like children in your own classroom.

As you examine each paper, look for a pattern of errors; then check your findings by using the error pattern yourself with the examples provided. In later chapters you will learn if your observations are accurate, and you will get feedback on your own suggestions for helping the child.

Be careful not to decide on the error pattern too quickly. (Children often make hasty decisions and, as a result, adopt the kind of erroneous procedures presented in this book.) When you think you see the pattern, verify your hypothesis by looking at the other examples on the child's paper.

If this book is to help you with your teaching of elementary school mathematics, you will need to "play the game." Take time to try out the error pattern before turning to another part of the book. Write out brief descriptions of instructional activities before moving ahead to see what suggestions are recorded later. Do not be content just to read about patterns of error; as a teacher you also learn by *doing,* and by *thinking* about what you are doing. Take time to respond in writing in the designated places.

Additional children's papers and a key are provided in Appendix A where you can further test your ability to identify error patterns. The tendency for children to adopt such patterns applies to more than arithmetical computation, so you may also want to study Appendix B where error patterns from other areas of mathematics are presented.

Error Pattern A-W-1

Examine Mike's work carefully. Can you find the error pattern he has followed?

Name *Mike*

A.
```
  7 4
+ 5 6
-----
1 2 1 0
```

B.
```
  3 5
+ 9 2
-----
1 2 7
```

C.
```
  6 7
+ 1 8
-----
7 1 5
```

D.
```
  5 6
+ 9 7
-----
1 4 1 3
```

Did you find his error pattern? Check yourself by using his error pattern to compute these examples.

E.
```
  4 3
+ 6 5
-----
```

F.
```
  8 8
+ 3 9
-----
```

Next, turn to pattern A-W-1 on page 96 to see if you were able to identify the error pattern. Why might Make or any student use such an erroneous computation procedure?

Error Pattern A-W-2

What error pattern is Mary following in her written work?

Name *Mary*

A.

$$432 \\ + 265 \\ \overline{697}$$

B.

$$\overset{1}{}74 \\ + 43 \\ \overline{18}$$

C.

$$38\overset{4}{5} \\ + 667 \\ \overline{9116}$$

D.

$$5\overset{\circ}{6}\overset{\circ}{3} \\ + 545 \\ \overline{118}$$

Check to see if you found Mary's pattern by using her erroneous procedure to compute these examples.

E.

$$254 \\ + 535 \\ \overline{}$$

F.

$$618 \\ + 782 \\ \overline{}$$

Next, turn to page 97 to see if you were able to identify Mary's error pattern. Why might Mary or any child use such a procedure?

Error Pattern A-W-3

Carol gets some correct answers, but she seems to miss many of the easiest examples. See if you can find her error pattern.

Name *Carol*

A.
```
  4 6
+    3
  1 3
```

B.
```
  1 8
+ 3 0
  4 8
```

C.
```
    8
+ 1 6
  1 5
```

D.
```
  4 2
+ 5 6
  9 8
```

E.
```
  8 5
+    6
  1 9
```

Use Carol's procedure for these examples to make sure you found her error pattern.

F.
```
  2 6
+    3
```

G.
```
  6 0
+ 2 4
```

H.
```
  7 4
+    5
```

When you complete examples F, G, and H, turn to page 98. Why might Carol be using such a procedure?

Error Pattern A-W-4

Can you find Dorothy's pattern of errors?

Name *Dorothy*

A.
$$
\begin{array}{r}
{}^{1}7\ 5 \\
+\quad 8 \\
\hline
1\ 6\ 3
\end{array}
$$

B.
$$
\begin{array}{r}
{}^{1}6\ 7 \\
+\quad 4 \\
\hline
1\ 1\ 1
\end{array}
$$

C.
$$
\begin{array}{r}
{}^{1}8\ 4 \\
+\quad 9 \\
\hline
1\ 8\ 3
\end{array}
$$

D.
$$
\begin{array}{r}
{}^{1}5\ 9 \\
6 \\
\hline
1\ 2\ 5
\end{array}
$$

Did you find the pattern? Make sure by using the error pattern to compute these examples.

E.
$$
\begin{array}{r}
4\ 6 \\
+\quad 8 \\
\hline
\end{array}
$$

F.
$$
\begin{array}{r}
9\ 8 \\
+\quad 3 \\
\hline
\end{array}
$$

When you complete examples E and F, turn to page 99 and see if you identified the pattern correctly. Why might Dorothy be using such a procedure?

Children can be observed using other error patterns while adding whole numbers. Practice your own diagnostic skills by identifying the patterns shown in Appendix A.

Error Pattern S-W-1

Look carefully at Cheryl's written work. What error pattern has she followed?
(**Note:** Cheryl used a different procedure for one of the examples.)

Name *Cheryl*

A. $\begin{array}{r} 3\,2 \\ -\ 1\,6 \\ \hline 1\,6 \end{array}$ B. $\begin{array}{r} 2\,4\,5 \\ -\ 1\,3\,7 \\ \hline 1\,1\,2 \end{array}$ C. $\begin{array}{r} 5\,2\,4 \\ -\ 2\,9\,8 \\ \hline 3\,7\,4 \end{array}$ D. $\begin{array}{r} 1\,3\,5 \\ -\ 6\,7 \\ \hline 1\,3\,2 \end{array}$

If you found the error pattern, check yourself by using that pattern to compute these examples.

E. $\begin{array}{r} 4\,5\,8 \\ -\ 3\,7\,2 \\ \hline \end{array}$ F. $\begin{array}{r} 2\,4\,1 \\ -\ 9\,6 \\ \hline \end{array}$

Now turn to pattern S-W-1 on page 100 to see if you were able to identify the error pattern. Why might a child use such a computational procedure?

Error Pattern S-W-2

Look over George's paper carefully. Can you find the error pattern he is using?

Name *George*

A.
$$
\begin{array}{r} 1\,\overset{8}{\cancel{9}}'7 \\ -\ 4\ 3 \\ \hline 1414 \end{array}
$$

B.
$$
\begin{array}{r} 1\,\overset{6}{\cancel{7}}'6 \\ -\ 2\ 3 \\ \hline 1\ 413 \end{array}
$$

C.
$$
\begin{array}{r} 3\,\overset{7}{\cancel{8}}4 \\ -\ 5\ 9 \\ \hline 3\ 25 \end{array}
$$

Did you find the error pattern? Check yourself by using George's error pattern to compute examples D and E.

E.
$$
\begin{array}{r} 2\ 7\ 3 \\ -\ 3\ 8 \\ \hline \end{array}
$$

F.
$$
\begin{array}{r} 2\ 8\ 5 \\ -\ 6\ 3 \\ \hline \end{array}
$$

If you have completed examples D and E, turn to page 101 to learn if you identified George's error pattern correctly. What instructional procedures might you use to help George or any other student with this problem?

Error Pattern S-W-3

Donna gets many incorrect answers when she subtracts. Can you find a pattern of errors?

Name *Donna*

A.
$$
\begin{array}{r}
1\,4\,7 \\
-\ \ 2\,0 \\
\hline
1\,2\,0
\end{array}
$$
B.
$$
\begin{array}{r}
6\,2\,4 \\
-3\,2\,3 \\
\hline
3\,0\,1
\end{array}
$$
C.
$$
\begin{array}{r}
5\,2\,7 \\
-3\,0\,4 \\
\hline
2\,0\,3
\end{array}
$$
D.
$$
\begin{array}{r}
8\,0\,5 \\
-2\,0\,1 \\
\hline
6\,0\,4
\end{array}
$$

Use Donna's error pattern to complete these examples.

E.
$$
\begin{array}{r}
4\,4\,6 \\
-3\,0\,2 \\
\hline
\end{array}
$$
F.
$$
\begin{array}{r}
7\,6\,0 \\
-2\,3\,0 \\
\hline
\end{array}
$$

After examples E and F are completed, turn to page 102 and see if you actually found Donna's pattern of errors. What might have caused Donna to begin such a procedure?

Error Pattern S-W-4

Barbara seemed to be doing well with subtraction until recently. Can you find a pattern of errors in her work?

Name *Barbara*

A.
$$
\begin{array}{r}
6\ \overset{8}{\cancel{7}}\ \overset{1}{\cancel{8}} \\
-2\ 4\ 8 \\
\hline
4\ 4\ 5
\end{array}
$$

B.
$$
\begin{array}{r}
\overset{2}{\cancel{3}}\ 2\ 5 \\
-1\ 5\ 1 \\
\hline
1\ 7\ 4
\end{array}
$$

C.
$$
\begin{array}{r}
\overset{5}{\cancel{7}}\ \overset{1}{2}\ 6 \\
-3\ 4\ 9 \\
\hline
2\ 8\ 7
\end{array}
$$

D.
$$
\begin{array}{r}
\overset{2}{\cancel{4}}\ \overset{1}{3}\ 4 \\
-2\ 7\ 6 \\
\hline
6\ 8
\end{array}
$$

To make sure you found the pattern, use Barbara's procedure to complete these examples.

E.
$$
\begin{array}{r}
4\ 3\ 6 \\
-\ 1\ 7\ 2 \\
\hline
\end{array}
$$

F.
$$
\begin{array}{r}
6\ 2\ 5 \\
-\ 3\ 4\ 8 \\
\hline
\end{array}
$$

After you complete examples E and F, turn to page 103 to see if you found Barbara's error pattern. What might have caused Barbara to begin using such a procedure?

Error Pattern S-W-5

Sam is having difficulty subtracting whole numbers. Do you find erroneous patterns in his work? He may be making more than one.

Name _Sam_

A.
```
  3 4
  3̷ 5̷
- 2 1
─────
  1 3
```

B.
```
  3 3
  2̷ 4 0
- 2 0 5
───────
1 3 0
```

C.
```
    4 5
  8̷ 6̷ 3
- 3 4 1
───────
1 1 2
```

D.
```
  5 6
  6̷ 7̷ 0
- 4 4 3
───────
1 2 0
```

In order to check your findings, use Sam's error patterns to compute examples E and F.

E.
```
  3 8 5
- 3 2 2
───────
```

F.
```
  6 4 0
- 6 2 6
───────
```

After you complete examples E and F, turn to page 104 to see if you found Sam's procedures. How would you help Sam or any child using procedures such as these?

Children can be observed using other error patterns while subtracting whole numbers. Practice your own diagnostic skills by identifying the patterns shown in Appendix A.

Error Pattern M-W-1

Examine Bob's written work carefully. What error pattern has he adopted?

Name _Bob_

A.
$$
\begin{array}{r}
\overset{2}{4}\,6 \\
\times\ \ 2\,4 \\
\hline
1\,8\,4 \\
1\,0\,2 \\
\hline
1\,2\,0\,4
\end{array}
$$

B.
$$
\begin{array}{r}
{}^{1}7\,6 \\
\times\ \ 3\,2 \\
\hline
1\,5\,2 \\
2\,2\,8 \\
\hline
2\,4\,3\,2
\end{array}
$$

C.
$$
\begin{array}{r}
\overset{5}{4}\,8 \\
\times\ \ 5\,7 \\
\hline
3\,3\,6 \\
2\,5\,0 \\
\hline
2\,8\,3\,6
\end{array}
$$

Were you able to identify Bob's error pattern? Check yourself by using the error pattern to compute examples D and E.

D.
$$
\begin{array}{r}
9\,8 \\
\times\ 5\,6 \\
\hline
\end{array}
$$

E.
$$
\begin{array}{r}
8\,6 \\
\times\ 4\,5 \\
\hline
\end{array}
$$

When you complete examples D and E, turn to page 104 to see if you identified Bob's error pattern correctly. What instructional procedures might you use to help Bob or another student with this problem?

Error Pattern M-W-2

Many of Bill's answers are not correct. Can you find an error pattern?

Name *Bill*

A.
$$\begin{array}{r} 3\ 4 \\ \times\ 2 \\ \hline 6\ 8 \end{array}$$

B.
$$\begin{array}{r} \overset{2}{2}\ 7 \\ \times\ 4 \\ \hline 8\ 8 \end{array}$$

C.
$$\begin{array}{r} \overset{1}{1}\ 8 \\ \times\ 3 \\ \hline 3\ 4 \end{array}$$

D.
$$\begin{array}{r} \overset{1}{2}\ 4 \\ \times\ 4 \\ \hline 8\ 6 \end{array}$$

Check yourself by using the error pattern you have observed to complete examples E and F.

E.
$$\begin{array}{r} 3\ 5 \\ \times\ 3 \\ \hline \end{array}$$

F.
$$\begin{array}{r} 2\ 8 \\ \times\ 4 \\ \hline \end{array}$$

Turn to page 105 and see if you correctly identified Bill's error pattern. How might you help Bill or other children who have adopted this procedure?

Error Pattern M-W-3

Joe's paper illustrates a common error pattern. Can you find it?

Name _Joe_

A.
$$\begin{array}{r} \overset{2}{2}\,7 \\ \times\,4 \\ \hline 1\,6\,8 \end{array}$$

B.
$$\begin{array}{r} \overset{2}{3}\,4 \\ \times\,6 \\ \hline 3\,0\,4 \end{array}$$

C.
$$\begin{array}{r} \overset{3}{4}\,5 \\ \times\,7 \\ \hline 4\,9\,5 \end{array}$$

When you think you have found Joe's error pattern, use his pattern to complete these examples.

D.
$$\begin{array}{r} 6\,8 \\ \times\quad 5 \\ \hline \end{array}$$

E.
$$\begin{array}{r} 2\,9 \\ \times\quad 3 \\ \hline \end{array}$$

After you finish examples D and E, turn to page 106 to see if you were able to identify the error pattern. What could possibly have caused Joe to learn such a procedure?

Error Pattern M-W-4

Doug seems to multiply correctly by a one-digit multiplier, but he is having trouble with two- and three-digit multipliers. Can you find this error pattern?

Name *Doug*

A.
$$\begin{array}{r} \overset{1}{3\,1\,3} \\ \times\ \ 4 \\ \hline 1\,2\,5\,2 \end{array}$$

B.
$$\begin{array}{r} 2\,1\,0 \\ \times\,1\ 5 \\ \hline 2\,1\,0 \end{array}$$

C.
$$\begin{array}{r} \overset{1}{5\,2\,4} \\ \times\,3\ 4 \\ \hline 1\,5\,7\,6 \end{array}$$

D.
$$\begin{array}{r} \overset{1}{4\,3\,3} \\ \times\,2\,2\,6 \\ \hline 8\,7\,8 \end{array}$$

Did you find his pattern? Check yourself by using Doug's error pattern to complete examples E and F.

E.
$$\begin{array}{r} 6\,2\,1 \\ \times\ \ \ 2\,3 \\ \hline \end{array}$$

F.
$$\begin{array}{r} 5\,1\,7 \\ \times\ 4\,6\,3 \\ \hline \end{array}$$

After examples E and F are completed, turn to page 107 to learn if you have correctly identified Doug's procedure. What instruction might you initiate with Doug or any child using such a procedure?

> You may observe children using other error patterns while multiplying whole numbers. Practice your own diagnostic skills by identifying the patterns shown in Appendix A.

Error Pattern D-W-1

Look very carefully at Jim's written work. What erroneous procedure has he used?

Name *Jim*

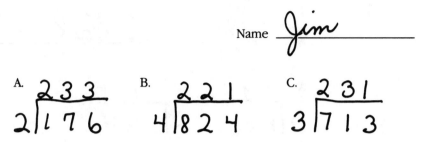

A. 233
2|176

B. 221
4|824

C. 231
3|713

Did you find the erroneous procedure? Check yourself by using Jim's procedure to compute examples D and E.

D.

3|639

E.

4|518

After completing examples D and E, turn to pattern D-W-1 on page 108 and learn if you correctly identified Jim's error pattern. What instructional procedures might you use to help Jim or any other student using this procedure?

Error Pattern D-W-2

Look carefully at Gail's written work. Can you find the error pattern she has followed?

Name *Gail*

A.
$$
\begin{array}{r}
4\ 4 \\
2\overline{\smash{)}8\ 8} \\
8 \\
\hline
8 \\
8 \\
\hline
\end{array}
$$

B.
$$
\begin{array}{r}
1\ 4 \\
4\overline{\smash{)}1\ 6\ 4} \\
1\ 6 \\
\hline
4 \\
4 \\
\hline
\end{array}
$$

C.
$$
\begin{array}{r}
6\ 7 \\
3\overline{\smash{)}2\ 2\ 8} \\
2\ 1 \\
\hline
1\ 8 \\
1\ 8 \\
\hline
\end{array}
$$

D.
$$
\begin{array}{r}
3\ 9 \\
5\overline{\smash{)}4\ 6\ 5} \\
4\ 5 \\
\hline
1\ 5 \\
1\ 5 \\
\hline
\end{array}
$$

Did you find the incorrect procedure? Check yourself by using the error pattern to compute these examples.

E.
$$3\overline{\smash{)}7\ 5}$$

F.
$$6\overline{\smash{)}5\ 1\ 6}$$

Next, turn to page 108 to see if you identified the error pattern correctly. Why might Gail or any child adopt such an incorrect procedure?

Error Pattern D-W-3

John has been doing well with much of his work in division, but he is having difficulty now. Can you find his error pattern?

Name _John_

A.
$$
\begin{array}{r}
6\,5\,R\,1 \\
7\overline{)4\ 5\ 6} \\
4\ 2 \\
\hline
3\ 6 \\
3\ 5 \\
\hline
1
\end{array}
$$

B.
$$
\begin{array}{r}
9\ \ 4\,R\,2 \\
6\overline{)5\ 4\ 2\ 6} \\
5\ 4 \\
\hline
2\ 6 \\
2\ 4 \\
\hline
2
\end{array}
$$

C.
$$
\begin{array}{r}
6\ 7\ R\,4 \\
8\overline{)4\ 8\ 6\ 0} \\
4\ 8 \\
\hline
6\ 0 \\
5\ 6 \\
\hline
4
\end{array}
$$

D.
$$
\begin{array}{r}
5\ 4\ R\,3 \\
8\overline{)4\ 0\ 3\ 5} \\
4\ 0 \\
\hline
3\ 5 \\
3\ 2 \\
\hline
3
\end{array}
$$

Try John's procedure with these examples.

E.
$$
9\overline{)2\ 7\ 2\ 1}
$$

F.
$$
6\overline{)4\ 2\ 5\ 0}
$$

After you complete examples E and F, turn to page 109 to learn if you correctly identified John's error pattern. Why might John be using such a procedure?

Error Pattern D-W-4

Anita seems to have difficulty with some division problems but she solves other problems correctly. Can you find her pattern of error?

Name *Anita*

A.
$$5\ 0_R4$$
$$5\overline{)2\ 5\ 4}$$
$$\underline{2\ 5\ 0}$$
$$4$$

B.
$$5\ 6\ 0_R6$$
$$9\overline{)4\ 5\ 6\ 0}$$
$$\underline{4\ 5\ 0\ 0}$$
$$6\ 0$$
$$\underline{5\ 4}$$
$$6$$

C.
$$7\ 3\ 0$$
$$8\overline{)5\ 8\ 4\ 0}$$
$$\underline{5\ 6\ 0\ 0}$$
$$2\ 4\ 0$$
$$\underline{2\ 4\ 0}$$

D.
$$3\ 7\ 0$$
$$7\overline{)2\ 1\ 4\ 9}$$
$$\underline{2\ 1\ 0\ 0}$$
$$4\ 9$$
$$\underline{4\ 9}$$

Did you find Anita's procedure? Check yourself by using her pattern to complete these examples.

E.
$$6\overline{)4\ 8\ 1\ 8}$$

F.
$$7\overline{)3\ 5\ 2\ 5}$$

When examples E and F are completed, turn to page 110 and learn if you found Anita's error pattern. How might you help Anita or any other child using such a procedure?

You may observe children using other error patterns while dividing whole numbers. Practice your own diagnostic skills by identifying the pattern shown in Appendix A.

Error Pattern E-F-1

Greg frequently makes errors when attempting to change a fraction to lower terms. What procedure is he using?

Name *Greg*

A. $\dfrac{1\ 9}{9\ 5} = \dfrac{1}{5}$ B. $\dfrac{1\ 3}{3\ 9} = \dfrac{1}{9}$

C. $\dfrac{1\ 8}{8\ 1} = \dfrac{1}{1}$ D. $\dfrac{1\ 2}{2\ 4} = \dfrac{1}{4}$

Determine if you correctly identified Greg's error pattern by using his procedure to change examples E and F to lower terms.

E. $\dfrac{16}{64} =$ F. $\dfrac{14}{42} =$

Now, turn to page 111 and see if you identified the error pattern correctly. Why might Greg be using such an incorrect procedure?

Error Pattern E-F-2

Jill determined the simplest terms for each given fraction, some already in simplest terms and some not. What procedure did she use?

Name _____Jill_____

A. $\dfrac{4}{9} = \dfrac{2}{3}$

B. $\dfrac{3}{9} = \dfrac{1}{3}$

C. $\dfrac{3}{8} = \dfrac{1}{4}$

D. $\dfrac{4}{8} = \dfrac{2}{4}$

Find out if you correctly identified Jill's procedure by using her error pattern to complete examples E and F.

E. $\dfrac{3}{4} =$

F. $\dfrac{2}{8} =$

Next, turn to page 112 where Jill's pattern is described. How might you help Jill or others using such a pattern?

Error Pattern E-F-3

Sue tried to change each fraction to lowest or simplest terms, but her results are quite unreasonable. Can you find her error pattern?

Name ___Sue___

A. $\dfrac{4}{8} = \dfrac{2}{8}$ B. $\dfrac{6}{8} = \dfrac{1}{8}$ C. $\dfrac{2}{4} = \dfrac{2}{4}$

D. $\dfrac{7}{7} = \dfrac{1}{7}$ E. $\dfrac{4}{6} = \dfrac{1}{6}$ F. $\dfrac{9}{3} = \dfrac{3}{9}$

Use Sue's procedures with these fractions to learn if you found her pattern.

G. $\dfrac{3}{6} =$ H. $\dfrac{6}{4} =$

After examples G and H are completed, turn to page 113 to see if you found Sue's error pattern. How would you help Sue or any child using such a procedure?

Error Pattern A-F-1

Can you find Robbie's pattern of error?

Name _Robbie_

A. $\frac{4}{5} + \frac{2}{3} = \frac{6}{8}$ B. $\frac{1}{4} + \frac{2}{3} = \frac{3}{7}$

C. $\frac{7}{8} + \frac{5}{6} = \frac{12}{14}$ D. $\frac{3}{7} + \frac{1}{2} = \frac{4}{9}$

Did you find the pattern? Make sure by using the pattern to compute these examples.

E. $\frac{3}{4} + \frac{1}{5} =$ F. $\frac{2}{3} + \frac{5}{6} =$

When you complete examples E and F, turn to page 114 and see if you identified the pattern correctly. Why might Robbie be using such a procedure? Can you think of a situation in which Robbie's procedure would actually be correct for the total needed? Think about sports.

Error Pattern A-F-2

What error pattern has Dave adopted?

Name *Dave*

A. $$\frac{3}{4} + \frac{2}{3} = \frac{5}{12}$$ B. $$\frac{6}{8} + \frac{1}{3} = \frac{7}{24}$$

C. $$\frac{2}{3} + \frac{5}{6} = \frac{7}{18}$$ D. $$\frac{3}{5} + \frac{2}{3} = \frac{5}{15} = \frac{1}{3}$$

Did you find Dave's error pattern? Check yourself by using his error pattern to compute examples E. and F.

E. $$\frac{1}{3} + \frac{3}{5} =$$ F. $$\frac{3}{8} + \frac{4}{5} =$$

When you complete examples E and F, turn to page 114 and see if you identified Dave's error pattern correctly. What instructional procedures could you use to help Dave or another student who computes in this way?

Error Pattern A-F-3

Allen is having difficulty with addition of unlike fractions. Is there a pattern of errors in his work?

Name *Allen*

A.
$$\frac{1}{4} + \frac{2}{3} = \frac{6}{7} + \frac{4}{7} = \frac{10}{7}$$

B.
$$\frac{1}{3} + \frac{3}{5} = \frac{15}{8} + \frac{3}{8} = \frac{18}{8}$$

C.
$$\frac{2}{3} + \frac{1}{2} = \frac{2}{5} + \frac{6}{5} = \frac{8}{5}$$

To find out if you identified Allen's error pattern, use his erroneous procedure to complete these examples.

D.
$$\frac{1}{4} + \frac{1}{5} =$$

E.
$$\frac{2}{5} + \frac{1}{2} =$$

When you complete examples D and E, turn to page 115 and see if you actually found Allen's pattern of errors. What might have caused Allen to begin using such a senseless procedure?

Error Pattern A-F-4

Robin is also having difficulty. What is she doing?

Name _Robin_

A.
$$\frac{1}{2} = \frac{1}{4}$$
$$+\frac{1}{4} = \frac{1}{4}$$
$$\frac{2}{4}$$

B.
$$\frac{2}{5} = \frac{2}{10}$$
$$+\frac{1}{2} = \frac{1}{10}$$
$$\frac{3}{10}$$

C.
$$\frac{3}{5} = \frac{3}{15}$$
$$+\frac{1}{3} = \frac{1}{15}$$
$$\frac{4}{15}$$

Did you determine what Robin is doing? Make sure by using her procedure to compute examples D and E.

D.
$$\frac{3}{4}$$
$$+\frac{1}{2}$$

E.
$$\frac{4}{5}$$
$$+\frac{1}{4}$$

When you complete examples D and E, turn to page 116 and see if you identified the procedure correctly. Why might Robin have learned to compute in this way?

You may observe children using other error patterns while adding with fractions. Practice your own diagnostic skills by identifying the patterns shown in Appendix A.

Error Pattern S-F-1

Andrew did fairly well with addition of fractions and mixed numbers, but he seems to be having trouble with subtraction. Can you find a pattern or patterns in his work?

Name *Andrew*

A.
$$7\frac{1}{2}$$
$$-\ 3$$
$$\overline{4\frac{1}{2}}$$

B.
$$8\frac{1}{3}$$
$$-\ \frac{2}{3}$$
$$\overline{8\frac{1}{3}}$$

C.
$$6$$
$$-\ 1\frac{1}{4}$$
$$\overline{5\frac{1}{4}}$$

D.
$$3\frac{1}{4}$$
$$-\ 2\frac{3}{4}$$
$$\overline{1\frac{3}{4}}$$

Make sure you found Andrew's pattern or patterns by using his procedures to complete these examples.

E.
$$5\frac{1}{5}$$
$$-\ 3\frac{3}{5}$$

F.
$$1$$
$$-\ \frac{1}{3}$$

When you complete examples E and F, turn to page 117 to see if you found Andrew's procedures. Why might Andrew be using such procedures?

Error Pattern S-F-2

Look carefully at Chuck's written work. What error pattern has he followed?

Name _Chuck_

A. $8\frac{3}{4} - 6\frac{1}{8} = 2\frac{2}{4}$ B. $5\frac{3}{8} - 2\frac{2}{3} = 3\frac{1}{5}$

C. $9\frac{1}{5} - 1\frac{3}{8} = 8\frac{2}{3}$ D. $7\frac{3}{5} - 4\frac{7}{10} = 3\frac{5}{5}$

If you found Chuck's error pattern, check yourself by using his procedure to compute these examples.

E. $6\frac{2}{3} - 3\frac{1}{6} =$ F. $4\frac{5}{8} - 1\frac{3}{4} =$

Now turn to page 117 to see if you identified the error pattern correctly. Why might a child use such a computational procedure?

Error Pattern S-F-3

Ann is having difficulty. Can you determine the faulty procedure she is using?

Name _Ann_

A.
$$2\frac{3}{4} = 2\frac{11}{4}$$
$$-1\frac{1}{2} = 1\frac{3}{4}$$
$$\overline{\qquad 1\frac{8}{4}}$$

B.
$$11\frac{1}{6} = 11\frac{67}{48}$$
$$-3\frac{7}{8} = 3\frac{31}{48}$$
$$\overline{\qquad 8\frac{36}{48}}$$

C.
$$9\frac{1}{3} = 9\frac{28}{3}$$
$$-\frac{2}{3} = \frac{2}{3}$$
$$\overline{\qquad 9\frac{26}{3}}$$

When you have found Ann's procedure, check yourself by using her procedure to compute these examples.

D.
$$5\frac{3}{8}$$
$$-2\frac{1}{2}$$

E.
$$4\frac{1}{3}$$
$$-1\frac{4}{5}$$

Turn to page 118 and see if you found the pattern of error. Why might a child use a procedure like this?

> You may observe children using other error patterns while subtracting with fractions. Practice your own diagnostic skills by identifying the patterns shown in Appendix A.

Error Pattern M-F-1

Dan is having considerable difficulty. What error pattern is he following in his written work?

Name $\underline{\textit{Dan}}$

A.
$$\frac{4}{5} \times \frac{3}{4} = 166$$

B.
$$\frac{1}{2} \times \frac{3}{8} = 68$$

C.
$$\frac{2}{9} \times \frac{1}{5} = 100$$

D.
$$\frac{2}{3} \times \frac{4}{6} = 132$$

Use Dan's procedure for these examples to make sure you found his error pattern.

E.
$$\frac{3}{4} \times \frac{2}{3} =$$

F.
$$\frac{4}{9} \times \frac{2}{5} =$$

When you complete examples E and F, turn to page 119. Why might Dan be using such a procedure?

Error Pattern M-F-2

Grace gets many answers correct when computing with fractions, but she is having difficulty with multiplication examples. See if you can find her error pattern.

Name *Grace*

A. $\dfrac{3}{8} \times \dfrac{5}{6} = \dfrac{3}{8} \times \dfrac{6}{5} = \dfrac{18}{40}$

B. $\dfrac{2}{5} \times \dfrac{3}{4} = \dfrac{2}{5} \times \dfrac{4}{3} = \dfrac{8}{15}$

C. $\dfrac{4}{5} \times \dfrac{2}{3} = \dfrac{4}{5} \times \dfrac{3}{2} = \dfrac{12}{10}$

Use Grace's procedure for the following examples to make sure you found her error pattern.

D. $\dfrac{2}{3} \times \dfrac{3}{4} =$

E. $\dfrac{5}{7} \times \dfrac{3}{8} =$

When you complete examples D and E, turn to page 120. Why might Grace be using such a procedure?

Error Pattern M-F-3

Look carefully at Lynn's paper. Can you find the procedure she is following?

Name *Lynn*

A. $\dfrac{1}{8} \times 1 = \dfrac{1}{8}$ B. $\dfrac{2}{3} \times 3 = \dfrac{6}{9}$

C. $\dfrac{1}{4} \times 6 = \dfrac{6}{24}$ D. $\dfrac{4}{5} \times 2 = \dfrac{8}{10}$

Did you find the procedure? Check yourself by using her error pattern to compute these examples.

E. $\dfrac{3}{8} \times 4 =$ F. $\dfrac{5}{6} \times 2 =$

Now that you have completed examples E and F, turn to page 120 and verify your response. Why should someone like Lynn learn to compute in this way?

You can observe children using other error patterns while multiplying with fractions. Practice your own diagnostic skills by identifying the patterns shown in Appendix A.

Error Pattern D-F-1

Linda has difficulty when she tries to divide with fractions. What procedure is she using?

Name *Linda*

A. $\dfrac{4}{6} \div \dfrac{2}{2} = \dfrac{2}{3}$ B. $\dfrac{6}{8} \div \dfrac{2}{8} = \dfrac{3}{1}$

C. $\dfrac{6}{10} \div \dfrac{2}{4} = \dfrac{3}{2}$ D. $\dfrac{7}{5} \div \dfrac{3}{2} = \dfrac{2}{2}$

Find out if you correctly identified Linda's procedure by using her error pattern to complete examples E and F.

E. $\dfrac{4}{12} \div \dfrac{4}{4} =$ F. $\dfrac{13}{20} \div \dfrac{5}{6} =$

Next, turn to page 121 where Linda's pattern is described. How might you help Linda and others using such a procedure?

Error Pattern D-F-2

Consider Joyce's work. Can you determine what procedure she is using?

Name *Joyce*

A. $\dfrac{2}{3} \div \dfrac{3}{8} = \dfrac{3}{2} \times \dfrac{3}{8} = \dfrac{9}{16}$

B. $\dfrac{2}{5} \div \dfrac{1}{3} = \dfrac{5}{2} \times \dfrac{1}{3} = \dfrac{5}{6}$

C. $\dfrac{3}{4} \div \dfrac{1}{5} = \dfrac{4}{3} \times \dfrac{1}{5} = \dfrac{4}{15}$

Were you able to determine her procedure? Does her procedure or pattern produce the correct product? Use Joyce's procedure for these examples to make sure.

D. $\dfrac{5}{8} \div \dfrac{2}{3} =$

E. $\dfrac{1}{2} \div \dfrac{1}{4} =$

When you complete examples D and E, turn to page 122. Why might Joyce be using such a procedure?

> You can observe children using other error patterns while dividing with fractions. Practice your own diagnostic skills by identifying the patterns shown in Appendix A.

Error Pattern A-D-1

Examine Harold's work carefully. Can you find the error pattern he is following?

Name *Harold*

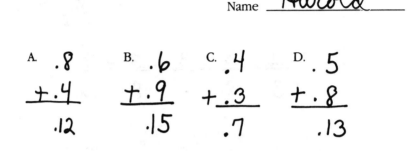

A.	B.	C.	D.
$.8$	$.6$	$.4$	$.5$
$+.4$	$+.9$	$+.3$	$+.8$
$.12$	$.15$	$.7$	$.13$

Did you find the pattern? Check yourself by using his error pattern to compute these examples.

E.	F.
$.3$	$.7$
$+.5$	$+.7$

After you complete examples E and F, turn to page 123 and verify your responses. What might have caused Harold to begin using such a procedure?

You can observe children using other error patterns while adding with decimals. Practice your own diagnostic skills by identifying the other patterns shown in Appendix A.

Error Pattern S-D-1

Les sometimes has difficulty when subtracting decimals. Can you determine his procedure?

Name *Les*

A.

$87-.31 = ?$

$$\begin{array}{r} 87 \\ -\ .31 \\ \hline 87.31 \end{array}$$

B.

$99.4-27.86 = ?$

$$\begin{array}{r} 99.4 \\ -27.86 \\ \hline 71.66 \end{array}$$

C.

$200-.65 = ?$

$$\begin{array}{r} 200 \\ -\ \ .65 \\ \hline 200.65 \end{array}$$

Find out if you correctly determined Les's error pattern by using his procedure to complete examples D and E.

D.

$60 - 1.35 = ?$

E.

$24.8 - 2.26 = ?$

If you completed examples D and E, turn to page 124 and check your responses. How might you help Les or others using such a procedure?

Error Pattern M-D-1

Marsha seems to have difficulty with some multiplication problems involving decimals but she solves other examples correctly. Can you find her pattern of error?

Name *Marsha*

A.
$$6.45$$
$$\times\ \ 3$$
$$19.35$$

B.
$$32.7$$
$$\times\ \ \ \ .5$$
$$16.35$$

C.
$$21.8$$
$$\times\ .4$$
$$87.2$$

D.
$$4.35$$
$$\times\ 2.3$$
$$1305$$
$$870$$
$$100.05$$

Did you find Marsha's procedure? Check yourself by using her pattern to complete these examples.

E.
$$40.5$$
$$\times\ \ \ \ .6$$

F.
$$6.7$$
$$\times\ \ 3$$

When examples E and F are complete, turn to page 124 and learn if you found Marsha's error pattern. How might you help Marsha or any other child using such a procedure?

Error Pattern D-D-1

Ted frequently makes errors when dividing decimals. Can you find this procedure?

Name *Ted*

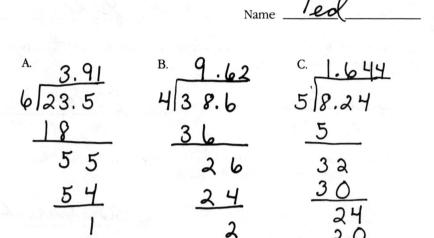

Use Ted's procedure with these examples to see if you found his error pattern.

D. $3\overline{)2.57}$

E. $.7\overline{)9.35}$

Now, turn to page 125 and see if you identified the error pattern correctly. Why might Ted be using such an incorrect procedure?

Error Pattern P-P-1

Sara correctly solves some percent problems, but many answers are incorrect. Can you find an error pattern?

Name *Sara*

A. On a test with 30 items, Mary worked 24 items correctly. What percent did she have correct?

$$\frac{24}{30} = \frac{x}{100}$$

Answer: *80 %*

B. Twelve students had perfect scores on a quiz. This is 40% of the class. How many students are in the class?

$$\frac{12}{40} = \frac{x}{100}$$

Answer: *30 students*

C. Jim correctly solved 88% of 50 test items. How many items did he have correct?

$$\frac{50}{88} = \frac{x}{100}$$

Answer: *57 items*

When you think you have found Sara's error pattern, use it to solve these problems.

D. Brad earned $400 during the summer and saved $240 from his earnings. What percent of his earnings did he save?

Answer: _____

E. Barbara received a gift of money on her birthday. She spent 80% of the money on a watch. The watch cost her $20. How much money did she receive as a birthday gift?

Answer: _____

F. The taffy sale brought in a total of $750, but 78% of this was used for expenses. How much money was used for expenses?

Answer: _____

Next, turn to page 126 to see if you identified the pattern correctly. Why might Sara or any child adopt such a procedure?

Error Pattern P-P-2

Steve is having difficulty solving percent questions. Can you find an error pattern in his paper?

Name _Steve_

A. What number is 30% of 180?

$$
\begin{array}{r}
1\,8\,0 \\
\times\ .3\,0 \\
\hline
0\,0\,0 \\
5\,4\,0 \\
\hline
5\,4.0\,0
\end{array}
$$

Answer: _54_

B. 15% of what number is 240?

$$
\begin{array}{r}
2\,4\,0 \\
\times\ .1\,5 \\
\hline
1\,2\,0\,0 \\
2\,4\,0 \\
\hline
3\,6.0\,0
\end{array}
$$

Answer: _36_

What percent of 40 is 28?

$$
\begin{array}{r}
4\,0 \\
\times\ .2\,8 \\
\hline
3\,2\,0 \\
8\,0 \\
\hline
1\,1.2\,0
\end{array}
$$

Answer: _11.2_

When you have found Steve's error pattern, use it to solve these examples.

D. What number is 80% of 54?   Answer: _____

E. Seventy is 14% of what number?   Answer: _____

F. What percent of 125 is 25?   Answer: _____

Now, turn to page 127 to learn if you found Steve's error pattern. How would you help a child using this procedure?

Error Pattern S-M-1

Margaret is having difficulty with computations that involve measurements. Can you find an error pattern in her work?

Name *Margaret*

A.
```
      4        1
    5 gallons, 1 quart
  - 1 gallon , 3 quarts
    3 gallons, 8 quarts
```

B.
```
    7        ′
  8 feet, 4 inches
 -3 feet, 9 inches
  4 feet, 5 inches
```

Check yourself by using Margaret's erroneous pattern to complete these examples.

C.
```
   6 yards, 1 foot
 - 2 yards, 2 feet
```

D.
```
   3 quarts, 1 cup
 - 1 quart, 3 cups
```

After you finish examples C and D, turn to page 128 to see if you accurately identified Margaret's error pattern. Why might Margaret or any other child use such a procedure?

In Appendix B, sample error patterns are shown for other areas of mathematics: numeration, problem solving, integers, algebra, and geometry. Practice your diagnostic skills by identifying the patterns illustrated.

Analyzing Error Patterns in Computation

In this chapter error patterns are described and analyzed. Accompanying many of the patterns is a discussion that focuses on reasons some children learn to compute with the particular error pattern. In many cases an erroneous computational procedure sometimes produces the correct answer, thereby confirming the validity of the pattern in the mind of the child. Evidences of purely mechanical procedures abound. Such procedures cannot be explained by the child through mathematical principles or with physical aids. Children who use mechanical procedures "push symbols around" whenever there are examples to be computed and right answers to be determined. Many children represented here have been introduced to the standard short-form algorithms too soon; some lack very basic understandings of numeration or the algorithm itself; others have become careless and confused. Each child has practiced an erroneous procedure.

As you read about error patterns you will have opportunities to suggest corrective instruction. These boys and girls may be in your classroom soon. What will you do to help them?

Keep in mind the suggestions included in Chapter 2 as you propose specific activities. Remember that some children may have special problems with language or in relating concepts, representations, and symbols. Does the child experiencing difficulty make sense out of manipulatives? Is the connection between manipulatives you use and the written procedure clear to the child?

Error Pattern A-W-1 *(used by Mike on page 56)*

Using the error patterns, examples E and F would be computed as shown.

E.
$$\begin{array}{r} 43 \\ +65 \\ \hline 108 \end{array}$$

F.
$$\begin{array}{r} 88 \\ +39 \\ \hline 1117 \end{array}$$

If your responses are the same as these, you were able to identify Mike's erroneous computational procedure. The ones were added and recorded, then the tens were added and recorded (or vice versa). The sum of the ones and the sum of the tens were each recorded without regard to place value in the sum. Note that Mike may have applied some knowledge of place value in his work with the two addends, *i.e.,* he may have treated the 88 in example F as 8 tens and 8 ones, and his answer as 11 tens and 17 ones. It is also true that Mike may have merely thought "8 plus 9 equals 17, and 8 plus 3 equals 11." I have found many students who think through such a problem in this way; some of these students also emphasize that you add 8 and 9 first because "you add the ones first."

Mike has the idea of adding ones with ones—possibly from work with bundles of sticks and single sticks. He apparently knows he should consider all of the single sticks together. He *may* know a rule for exchanging or regrouping ten single sticks for one bundle of ten, but, if he does, he has not applied the rule to these examples. Previous instruction may not have given adequate emphasis to the mechanics of recording sums.

Cox found, in her study of systematic errors among children in regular second- through sixth-grade classrooms, that 67 percent of the children who made systematic errors when adding two two-digit numbers with renaming made this particular error.[1]

If you were Mike's teacher, how would you help him? Describe two instructional activities that might correct the error pattern.

1. _____

_____

2. _____

_____

When you have completed your responses, turn to page 130 to see if your suggestions are among the alternatives described.

Error Pattern A-W-2 *(from Mary's paper on page 57)*

If you used Mary's error pattern, you completed examples E and F as they are shown.

E.
```
  2 5 4
+ 5 3 5
  7 8 9
```

F.
```
    3 2
  6 1 8
+ 7 8 2
1 1 1 2
```

This pattern is a reversal of the procedure used in the usual algorithm—without regard for place value. Addition is performed from left to right, and, when the sum of a column is ten or greater, the left figure is recorded and the right figure is placed above the next column to the right.

You probably noted that example A on page 57 and example E are correct. In these two examples, Mary's use of a left-to-right procedure was reinforced because she computed the correct sum. You may want to determine if Mary's left-to-right orientation is from reading instruction, especially if she is in a remedial program. She may need help in taking a "global look" at numerals to identify place values.

If you were Mary's teacher, what corrective procedures might you follow? Describe two instructional activities that might help Mary add correctly.

1. _____

   _____

2. _____

   _____

When your responses are complete, turn to page 131 and see if your suggestions are among the alternatives described.

Error Pattern A-W-3 *(from Carol's paper on page 58)*

If you found Carol's error pattern, your results are the same as the erroneous computation shown.

F.
$$\begin{array}{r} 2\ 6 \\ +\quad 3 \\ \hline 1\ 1 \end{array}$$

G.
$$\begin{array}{r} 6\ 0 \\ +2\ 4 \\ \hline 8\ 4 \end{array}$$

H.
$$\begin{array}{r} 7\ 4 \\ +\quad 5 \\ \hline 1\ 6 \end{array}$$

Carol misses examples in which one of the addends is written as a single digit. When working such examples, she adds the three digits as if they were all units. When both addends are two-digit numbers, she appears to add correctly. However, it is quite probable that Carol is not applying any knowledge of place value with either type of example. She may be merely adding units in every case. (When both addends are two-digit numbers, she adds units in straight columns. When one addend is a one-digit number, she adds the three digits along a curve.) If this is the way Carol is thinking, she will probably experience even more failure and frustration when she begins to add and subtract examples that require renaming.

　　How would you help Carol? Describe at least two instructional activities you believe would correct Carol's error pattern.

1. _____

_____

2. _____

_____

　　After you have described at least two activities, turn to page 133 and compare your suggestions with those listed there.

Error Pattern A-W-4 (from Dorothy's paper on page 59)

Using Dorothy's error pattern, examples E and F would be computed as shown.

E.
$$
\begin{array}{r}
\overset{1}{4}\,6 \\
+\ \ 8 \\
\hline
1\,3\,4
\end{array}
$$

F.
$$
\begin{array}{r}
\overset{1}{9}\,8 \\
+\ \ 3 \\
\hline
1\,3\,1
\end{array}
$$

　　Dorothy is not having difficulty with her basic addition facts, but higher decade addition situations are confusing her. She tries to use the regular addition algorithm; however, when she adds the tens column she adds in the one-digit number again.

　　If Dorothy has been introduced to the multiplication algorithm, she may persist in seeing similar patterns for computation whenever numerals are arranged as she has seen them in multiplication examples. Changing operations when the arrangement of digits is similar is difficult for some children. An interview with Dorothy may help you determine if she really knows how to add problems like these. When working with a child who tends to carry over one situation into her perception of another, avoid extensive practice at a given time on any single procedure.

　　In her study of children adding a two-digit number and a one-digit number with renaming, Cox found that 11 percent of the children in regular classrooms who made systematic errors made this particular error. Furthermore, in special education classrooms 60 percent of the children with systematic errors had adopted this procedure.[2]

How would you help Dorothy? Describe at least two instructional activities that would help Dorothy replace her erroneous pattern with a correct procedure.

1. _____

_____

2. _____

_____

When both activities are described, turn to page 134 and compare your suggestions with those recorded there.

Error Pattern S-W-1 (from Cheryl's paper on page 60)

Using the error pattern, examples E and F would be computed as shown.

$$
\begin{array}{r}
\text{E.} \quad 4\ 5\ 8 \\
-3\ 7\ 2 \\
\hline
1\ 2\ 6
\end{array}
\qquad
\begin{array}{r}
\text{F.} \quad 2\ 4\ 1 \\
-\ \ 9\ 6 \\
\hline
2\ 5\ 5
\end{array}
$$

Did you identify the error pattern? As a general rule, the ones are subtracted and recorded, then the tens are subtracted and recorded, etc. Apparently Cheryl is considering each position (ones, tens, etc.) as a separate subtraction problem. In example E she probably did not think of the numbers 458 and 372, but only of 8 and 2, 5 and 7, and 4 and 3. Further, in subtracting single-digit numbers, she does not conceive of the upper figure (minuend) as the number in a set and the lower figure (subtrahend) as the number in a subset. When subtracting ones Cheryl may think of the larger of the two numbers as the number of the set, and the smaller as the number to be removed from the set. Or she may merely compare the two single-digit numbers much as she would match sets one to one or place rods side by side to find a difference. In example F she would think "1 and 6, the difference is 5." She uses the same procedure when subtracting tens and hundreds. Cheryl may have merely overgeneralized commutativity for addition, and assumed that subtraction is also commutative.

Note that example A on page 60 is correct. This example includes much smaller numbers than the other examples. It may be that Cheryl counted from

16 to 32, or she may have used some kind of number line. If she did think of 32 as "20 plus 12" in order to subtract, it may be that she applies renaming procedures only to smaller numbers that she can somehow conceptualize, but she breaks up larger numbers in the manner described above.

Has Cheryl heard rules in the classroom or at home that she is applying in her own way? Perhaps she has heard, "Always subtract the little number from the big one" and "Stay in the column when you subtract."

Children frequently adopt this error pattern. For children in regular classrooms who were subtracting a two-digit number from a two-digit number with renaming, Cox found that 83 percent of the children with systematic errors used this particular procedure.[3]

If you were Cheryl's teacher, what would you do? Describe two instructional activities that might help Cheryl correct the error pattern.

1. _____

_____

2. _____

_____

After you have finished writing your responses, turn to page 135 to see if your suggestions are among the alternatives described.

Error Pattern S-W-2 *(from George's paper on page 61)*

Did you identify the error pattern George used?

D.
$$2\ \overset{6}{\cancel{7}}\ ^{1}3$$
$$-\ 3\ 8$$
$$\overline{2\ 3\ 5}$$

E.
$$2\ \overset{7}{\cancel{8}}\ ^{1}5$$
$$-\ 6\ 3$$
$$\overline{2\ 1\ 1\ 2}$$

George has learned to "borrow" or rename in subtraction. In fact, he renames whether he needs to or not. It is possible that George would be able to rename one ten as ten ones, and it is also possible that he could interpret the answer (in example E) as 12 ones, 1 ten, and 2 hundreds. But his final answer does not take account of conventional place-value notation.

You have no doubt observed that the answer to example D is correct. George's procedure is correct for a subtraction example that requires renaming tens as ones. However, George has overgeneralized. He does not distinguish between examples that require renaming and examples that do not require renaming. The fact that some of his answers are correct may reinforce his perception that the procedure he is using is appropriate for all subtraction examples.

Many children rename the minuend when it is unnecessary. In Cox's study, when children in regular classrooms subtracted a two-digit number from a two-digit number with no renaming, 75 percent of the children who used a systematic erroneous procedure did rename the minuend although it was inappropriate to do so.[4]

How would you help George with his problem? Describe two instructional activities that would help George replace his error pattern with a correct computational procedure.

1. _____

_____

2. _____

_____

After you complete your two descriptions, turn to page 136 and compare what you have written with the suggestions presented there.

Error Pattern S-W-3 *(from Donna's paper on page 62)*

Did you find Donna's error pattern?

$$
\begin{array}{r}
\text{E.} \quad 446 \\
-302 \\
\hline
104
\end{array}
\qquad
\begin{array}{r}
\text{F.} \quad 760 \\
-230 \\
\hline
530
\end{array}
$$

Although Donna uses the subtraction fact $0 - 0 = 0$ correctly, she consistently writes "0" for the missing addend whenever the known addend (subtrahend) is zero. She regularly misses nine of the 100 basic subtraction facts because of this one difficulty.

We ought to be able to help Donna with a problem of this sort. How would you help her? Describe two instructional activities you think would enable Donna to subtract correctly.

1. _____

   _____

2. _____

   _____

Did you describe at least two activities? (It is important to have more than one possible instructional procedure in mind when working remedially with a child.) If so, turn to page 137 and compare your suggestions with the suggestions described there.

Error Pattern S-W-4 *(from Barbara's paper on page 63)*

Did you find Barbara's error pattern?

E.
$$
\begin{array}{r}
\overset{3}{\cancel{4}}\overset{1}{3}\,6 \\
-\;1\;7\;2 \\
\hline
2\;6\;4
\end{array}
$$

F.
$$
\begin{array}{r}
\overset{4}{\cancel{6}}\overset{1}{2}\overset{1}{5} \\
-\;3\;4\;8 \\
\hline
1\;8\;7
\end{array}
$$

Barbara appeared to be doing well with subtraction until recently, when renaming more than once was introduced. She apparently had been thinking something like "Take 1 from 4 and put the 1 in front of the 3" (example E). Now she has extended this procedure so that, in example F, she thinks, "In order to subtract (*i.e.,* in order to use a simple subtraction fact), I need a 1 in front of the 5 and a 1 in front of the 2. Take *two* 1's from the 6 ...." Note that if Barbara had not been showing her work with crutches of some sort, it would have been much more difficult to find the pattern.

Help is needed, and promptly—before she reinforces her error pattern with further practice. How would you help her? Describe two instructional activities you think would make it possible for Barbara to subtract correctly, even in examples such as these.

1. _____

   _____

2. _____

   _____

When you have described two instructional activities, turn to page 139 and compare your suggestions with those listed there.

Error Pattern S-W-5 *(from Sam's paper on page 64)*

Did you find Sam's error patterns?

E.
$$\overset{\overset{4}{\cancel{8}}}{\,}\,\overset{7}{\cancel{8}}\,5$$
$$-3\,2\,2$$
$$\overline{1\,5\,3}$$

F.
$$\overset{7}{\cancel{6}}\,\overset{3}{\cancel{4}}\,0$$
$$-6\,2\,6$$
$$\overline{1\,1\,0}$$

Whenever the digits in the minuend and the subtrahend are the same, Sam borrows "so he will be able to subtract." However, his procedure is simply "Take one from here and add it here." It is not meaningful renaming. You may also have noted that Sam has difficulty with zeros in the minuend. Rather than regrouping so he can subtract, he merely records a zero.

How would you help Sam if you had the opportunity? Describe at least two instructional activities that you believe would help Sam subtract whole numbers correctly.

1. _____

_____

2. _____

_____

After you have written both descriptions, turn to page 140 and see if your suggestions are among those listed there.

Error Pattern M-W-1 *(used by Bob on page 65)*

Did you find the error pattern Bob used?

Consider example E. When multiplying 5 ones times 6 ones, Bob recorded the 3 tens as a crutch above the 8 tens to remind him to add 3 tens to the product of 5 and 8 tens. However, the crutch recorded when multiplying by ones was *also* used when multiplying by tens.

D.
```
    ⁴9 8
     5 6
   5 8 8
   4 9 0
   5 4 8 8
```

E.
```
    ³8 6
     4 5
   4 3 0
   3 5 4
   3 9 7 0
```

Note that the answers to example B on page 65 and example D above are correct. Bob's error pattern may have gone undetected because he gets enough correct answers, enough positive reinforcement, to convince him that he is using a correct procedure. There may have been enough correct answers to cause Bob's busy teacher to conclude that Bob was merely careless. But Bob *is* consistently applying an erroneous procedure.

How would you help Bob with this problem? Describe two instructional activities that would help Bob replace this error pattern with a correct computational procedure.

1. _____

_____

2. _____

_____

When you have written descriptions of two instructional activities, turn to page 141 and compare what you have written with the suggestions presented there.

Error Pattern M-W-2 *(from Bill's paper on page 66)*

The following examples illustrate Bill's error pattern.

E.
```
    ¹3 5
  ×    3
     9 5
```

F.
```
    ³2 8
  ×    4
     8 2
```

Bill simply does not add the number of tens he records with a crutch. Perhaps he knows he should add, but don't be too sure. He forgets consistently!

How would you help Bill with a problem of this sort? Describe two activities you think would enable Bill to complete the multiplication correctly.

1. _____

   _____

2. _____

   _____

Are both activities described? Then turn to page 144 and compare what you have written with the suggestions listed there.

Error Pattern M-W-3 *(from Joe's paper on page 67)*

Did you find the error pattern Joe uses?

$$
\begin{array}{r}
\overset{4}{6}\,8 \\
\times\ \ 5 \\
\hline
5\,0\,0
\end{array}
\qquad
\begin{array}{r}
\overset{2}{2}\,9 \\
\times\ \ 3 \\
\hline
1\,2\,7
\end{array}
$$

D.   E.

Joe is using an erroneous procedure that is all too frequently adopted by children. He adds the number associated with the crutch *before* multiplying the tens figure, whereas the algorithm requires that the tens figure be multiplied first. In example D, he thought "6 plus 4 equals 10 and 5 times 10 equals 50" instead of "5 times 6 equals 30 and 30 plus 4 equals 34." It may be that when Joe learned the addition algorithm involving regrouping, his teacher reminded him repeatedly, "The first thing you do is to add the number you carry." Many teachers drill children on such a rule, and it is little wonder that children sometimes apply the rule in inappropriate contexts.

Cox found in her study that, for examples of this type, 43 percent of the children in regular classrooms who used an erroneous procedure systematically used this particular error pattern.[5]

The fact that children frequently use Joe's procedure does not lessen your obligation to help Joe multiply correctly. How would *you* help him? Describe two different instructional activities you think would enable Joe to replace his error pattern with a correct computational procedure.

1. _____

_____

2. _____

_____

When you have finished writing both descriptions, turn to page 145 and compare what you have written with the suggestions offered there.

Error Pattern M-W-4 *(from Doug's paper on page 68)*

Did you find Doug's error pattern? If so, you completed examples E and F as they are shown.

$$
\begin{array}{r}
\text{E.} \quad 6\,2\,1 \\
\times \quad 2\,3 \\
\hline
1\,2\,4\,3
\end{array}
\qquad
\begin{array}{r}
\text{F.} \quad 5\,\overset{2}{1}\,7 \\
\times \ 4\,6\,3 \\
\hline
2\,0\,8\,1
\end{array}
$$

The procedure used by Doug is a blend of the algorithm for multiplying by a one-digit multiplier and the conventional addition algorithm. Each column is approached as a separate multiplication; when the multiplicand has more digits than the multiplier, the left-most digit of the multiplier continues to be used.

You may meet Doug in your own classroom. How would you help him? Describe at least two instructional activities you believe would enable Doug to multiply by two- and three-digit numbers correctly.

1. _____

_____

2. _____

_____

After your descriptions are written, turn to page 146 and see if your suggestions are among those listed there.

Error Pattern D-W-1 *(used by Jim on page 69)*

Did you find the erroneous procedure Jim used?

D.
$$3\overline{)639} = 2\,/\,3$$

E.
$$4\overline{)518} = /\,42$$

Example D is correct and it does not give many clues to Jim's thinking. However, the fact that example D *is* correct is a reminder that erroneous procedures sometimes produce correct answers, thereby reinforcing the procedure as far as the student is concerned and making the error pattern more difficult for the teacher to identify.

Example E illustrates Jim's thinking more completely. Apparently Jim ignores place value in the dividend and quotient, and he thinks of each digit as "ones." Furthermore, he considers one digit of the dividend and the one-digit divisor as two numbers "to be divided." The greater of the two (whether the divisor or a digit within the dividend) is divided by the lesser and the result is recorded. Jim has probably learned something like "a smaller number goes into a larger number." Interestingly, the remainder is ignored.

Did you notice that no numerals were recorded below the dividend? This lack in itself is a sign of possible trouble. It may well be that someone (a teacher, a parent, a friend) tried to teach Jim the "short" division procedure.

How would you help Jim with his problem? Describe two instructional activities that you think would help Jim replace this erroneous procedure with a correct computational procedure.

1. _____

    _____

2. _____

    _____

After you have written descriptions of two appropriate instructional activities, compare your activities with the suggestions on page 148.

Error Pattern D-W-2 *(from Gail's paper on page 70)*

Using Gail's incorrect procedure, examples E and F would be computed as shown.

E.

$$\begin{array}{r} 52 \\ 3\overline{)75} \\ \underline{60} \\ 15 \\ \underline{15} \end{array}$$

F.

$$\begin{array}{r} 68 \\ 6\overline{)516} \\ \underline{480} \\ 36 \\ \underline{36} \end{array}$$

Did you find the error pattern? In the ones column Gail records the first quotient figure she determines, and in the tens column she records the second digit she determines. In other words, the answer is recorded right to left. In the usual algorithms for addition, subtraction, and multiplication of whole numbers, the answer is recorded right to left; perhaps Gail assumes it is appropriate to do the same with the division algorithm.

The fact that example A is correct illustrates again that correct answers are sometimes obtained with incorrect procedures, thereby positively reinforcing an error pattern.

It is quite probable that, for example E, Gail thinks "7 divided by 3" (or perhaps "3 times what number is 7") rather than "75 divided by 3." The quotient for a shortcut expression such as "7 divided by 3" would indeed be 2 units. Shortcuts in thinking and the standard algorithm for division may have been introduced too soon.

What would you do if you were Gail's teacher? What corrective steps might you take? Describe two instructional activities that you believe would help Gail correct the erroneous procedure.

1. _____

_____

2. _____

_____

If you have finished your responses, turn to page 150 and see if your suggestions are among the alternatives listed there.

Error Pattern D-W-3 (from John's paper on page 71)

If you found John's error pattern, you completed examples E and F as they are shown.

E.
$$9\overline{)2721} \quad \frac{3 \quad 2}{} R3$$

```
      3   2  R3
    _____
  9 | 2 7 2 1
    2 7
    ___
        2 1
        1 8
        ___
          3
```

F.
```
      7   8  R2
    _____
  6 | 4 2 5 0
    4 2
    ___
        5 0
        4 8
        ___
          2
```

John has difficulty with examples that include a zero in the tens place of the quotient. Whenever he brings down and cannot divide he brings down again, but without recording a zero to show that there are no tens. He may believe that "zero is nothing." Also, careless placement of figures in the quotient may contribute to John's problem.

Children commonly compute this way. Cox found that for children with systematic errors in regular classrooms who computed similar examples (but with three-digit dividends) 72 percent used this procedure. In special education classrooms 90 percent of those with systematic errors had adopted this error pattern.[6]

How would you help John? Describe at least two instructional activities you believe would help John correct his pattern of error.

1. _____

_____

2. _____

_____

When you have completed both descriptions, turn to page 153 and compare your suggestions with those listed there.

Error Pattern D-W-4 (from Anita's paper on page 73)

If Anita's procedure is used, examples E and F would be completed as shown.

E.

$$6\overline{)4\,8\,1\,8}$$ quotient 830
$$4\,8\,6\,0$$
$$1\,8$$
$$1\,8$$

F.

$$7\overline{)3\,5\,2\,5}$$ quotient $530\,r4$
$$3\,5\,0\,0$$
$$2\,5$$
$$2\,1$$
$$4$$

Like John, Anita is having difficulty with examples that include a zero in the tens place of the quotient. If she "brings down" and cannot divide, she "brings down" again; but she does not record a zero to show that there are no tens. However, she is careful to align her work in columns of hundreds, tens, etc., and the quotient is obviously incomplete as she finishes her computation. Therefore, a zero is inserted in the remaining position—the ones place.

The extensive use of zeros in the computation (*e.g.*, the 4800 in example E) sometimes helps a child if the child consciously multiplies in terms of the total values involved. In example E, 800 × 6 = 4800. However, Anita also indicated that 30 × 6 = 18, which is not true. It may be the case that zeros are being written all the way across just because it is the thing to do. An interview with Anita, letting her think out loud while computing, may be helpful in determining what is actually happening.

What would *you* do to help Anita? Describe at least two instructional activities you believe would help her correct the pattern of error.

1. _____

   _____

2. _____

   _____

When both activities have been described, turn to page 153 and compare your suggestions with those recorded there.

Error Pattern E-F-1 *(from Greg's paper on page 73)*

If Greg's procedure is used, examples E and F would be completed as follows.

E. $\dfrac{16}{64} = \dfrac{1}{4}$

F. $\dfrac{14}{42} = \dfrac{1}{2}$

If the same digit appears in both numerator and denominator, Greg uses a cancellation procedure similar to what he learned in another context. Though the procedure Greg uses is erroneous, examples A and E are both correct. Greg may find it difficult to believe his method is not satisfactory.

We ought to be able to help Greg. How would you help him? Describe two instructional activities you believe would help Greg correct his pattern of error.

1. _____

_____

2. _____

_____

After two activities have been described, turn to page 155 and compare what you have written with the suggestions listed there.

Error Pattern E-F-2 *(from Jill's paper on page 74)*

Did you find Jill's error pattern?

$$\text{E.} \quad \frac{3}{4} = \frac{1}{2} \qquad \text{F.} \quad \frac{2}{8} = \frac{1}{4}$$

Jill's computation appears almost random though, interestingly, several answers are correct. She obviously does not recognize which fractions are already in simplest terms. She explains her procedure as follows:

"4 goes to 2, and 9 goes to 3"
"3 goes to 1, and 9 goes to 3"

For examples E and F,

"3 goes to 1, and 4 goes to 2"
"2 goes to 1, and 8 goes to 4"

Jill simply associates a specific whole number with each given numerator or denominator. *All* 3's become 1's and *all* 4's become 2's when fractions are to be reduced or changed to simplest terms. This procedure is a very mechanical one, requiring no concept of a fraction; however, it does produce correct answers part of the time.

Concepts such as the equivalence of rational numbers develop *very slowly* over time. If students are taught computational procedures before their concepts of fraction and equivalent fractions are adequate, they are apt to experience difficulty.

Jill has a very real problem. How would you help her? Describe two instructional activities that would help Jill replace this erroneous procedure with a correct procedure.

1. _____

_____

2. _____

_____

After two activities have been described, turn to page 156 and compare what you have written with the suggestions listed there.

Error Pattern E-F-3 *(from Sue's paper on page 75)*

Did you correctly identify Sue's pattern of errors? Many teachers would assume she had responded randomly.

$$\text{G.} \quad \frac{3}{6} = \frac{2}{6} \qquad\qquad \text{H} \quad \frac{6}{4} = \frac{1}{6}$$

Sue considers the given numerator and denominator as two whole numbers, and divides the greater by the lesser to determine the new numerator (ignoring any remainder); then the greater of the two numbers is copied as the new denominator. Perhaps she has observed, in the fractions she has seen, that the denominator is usually the greater of the two numbers.

How would you help Sue? She is not unlike many other children who develop mechanistic and unreasonable procedures in arithmetic classes. Describe at least two instructional activities that you believe would help Sue learn to correctly change fractions to lowest or simplest terms.

1. _____

_____

2. _____

_____

When you have described at least two activities, turn to page 158 and compare your suggestions with those listed there.

Error Pattern A-F-1 (from Robbie's paper on page 76)

Did you find Robbie's error pattern?

$$\text{E.} \quad \frac{3}{4} + \frac{1}{5} = \frac{4}{9} \qquad \qquad \text{F.} \quad \frac{2}{3} + \frac{5}{6} = \frac{7}{9}$$

Robbie adds the numerators to get the numerator for the sum, then he adds the denominators to get the denominator for the sum. Lankford, in the research described in Chapter 1, found this procedure to be a "most prevalent practice."[7] It is likely that Robbie has already learned to multiply fractions, and he appears to be following a similar procedure for adding fractions.

How would you help Robbie? Describe two instructional activities that would help him replace his error pattern with a correct computational procedure.

1. _____

    _____

2. _____

    _____

After you complete your two descriptions, turn to page 160 and compare what you have written with the suggestions presented there.

Error Pattern A-F-2 (from Dave's paper on page 77)

Did you find the error pattern Dave is using?

$$\text{E.} \quad \frac{1}{3} + \frac{3}{5} = \frac{4}{15} \qquad \qquad \text{F.} \quad \frac{3}{8} + \frac{4}{5} = \frac{7}{40}$$

Dave may remember that you often have to multiply when adding fractions like these, although he does not apply any understanding of common denominators or renaming fractions. He merely adds the numerators to get the numerator for the sum, and multiplies the denominators to get the denominator for the sum.

How would you help Dave? Describe two instructional activities that you think would help Dave replace his error pattern with a correct computational procedure.

1. _____

_____

2. _____

_____

When you have finished describing both instructional activities, turn to page 161 and compare your suggestions with those listed there.

Error Pattern A-F-3 *(from Allen's paper on page 78)*

Did you find Allen's error pattern?

$$\text{D.} \quad \frac{1}{4} + \frac{1}{5} = \frac{5}{9} + \frac{4}{9} = \frac{9}{9}$$

$$\text{E.} \quad \frac{2}{5} + \frac{1}{2} = \frac{2}{7} + \frac{10}{7} = \frac{12}{7}$$

Allen has learned a very mechanical procedure for changing two fractions so they will have the same denominator. Someone may have tried to teach him a rather common shortcut that includes rules for adding and multiplying different numerators in an apparently arbitrary pattern. However, what Allen actually learned was a very difficult pattern of additions and multiplications, a mechanical procedure he uses to "get an answer" when required to do so by a teacher. He first adds the unlike denominators to get a common denominator, then he multiplies the numerator and denominator of the first fraction to get a numerator for the new second fraction. Similarly, he multiplies the numerator and denominator of the second fraction to get a numerator for the new first fraction. (Apparently, he adds like fractions correctly.) If Allen has been practicing addition of unlike fractions, it is this erroneous procedure he has been reinforcing.

Can you help Allen with a problem of this sort? Describe at least two instructional activities you think would enable Allen to add unlike fractions correctly.

1. _____

_____

2. _____

_____

Have you described at least two activities? Remember, it is important to have more than one possible instructional procedure in mind when working remedially with a child. If so, turn to page 162 and compare your suggestions with those noted there.

Error Pattern A-F-4 *(from Robin's paper on page 79)*

Did you find Robin's procedure?

D.
$$\frac{3}{4} = \frac{3}{4}$$
$$+\frac{1}{2} = \frac{1}{4}$$
$$\frac{4}{4}$$

E.
$$\frac{4}{5} = \frac{4}{20}$$
$$+\frac{1}{4} = \frac{1}{20}$$
$$\frac{5}{20}$$

Robin is able to determine the least common denominator and she uses it when changing two fractions so they will have the same denominator. However, she merely copies the original numerator. Apparently, Robin is able to add like fractions correctly.

How would you help Robin with her difficulty? Describe two instructional activities that you believe would help her learn to add unlike fractions correctly.

1. _____

_____

2. _____

_____

When you have described both activities, turn to page 164 and compare your suggestions with those listed there.

Error Pattern S-F-1 *(from Andrew's paper on page 80)*

Did you find Andrew's error patterns?

$$\text{E.} \quad 5\tfrac{1}{5} \\ -3\tfrac{3}{5} \\ \hline 2\tfrac{2}{5}$$

$$\text{F.} \quad 1 \\ -\tfrac{1}{3} \\ \hline 1\tfrac{1}{3}$$

In every case the whole numbers are subtracted as simple subtraction problems, perhaps even before attention is given to the column of common fractions. Where only one fraction appears in the problem (example F) the fraction is simply "brought down." If two fractions appear, Andrew records the difference between them, ignoring whether the subtrahend or the minuend is the greater of the two.

How would you help Andrew? Describe two instructional activities you think would help Andrew correct his erroneous procedures.

1. _____

   _____

2. _____

   _____

When both descriptions are completed, turn to page 165 and compare your suggestions with those listed there.

Error Pattern S-F-2 *(from Chuck's paper on page 81)*

Did you find Chuck's error pattern?

$$\text{E.} \quad 6\tfrac{2}{3} - 3\tfrac{1}{6} = 3\tfrac{1}{3}$$

$$\text{F.} \quad 4\tfrac{5}{8} - 1\tfrac{3}{4} = 3\tfrac{2}{4}$$

Chuck is subtracting by first finding the difference between the two whole numbers and recording that difference as the new whole number. He then finds the difference between the two numerators and records that difference as the new numerator. Finally, he finds the difference between the two denominators and records that number as the new denominator. The procedure is similar to addition as seen in Error Pattern A-F-1. However, children using this procedure for subtraction necessarily ignore the order of the minuend and the subtrahend.

Someone needs to come to Chuck's aid. How would you help him? Describe two instructional procedures you believe would help Chuck subtract correctly when given examples such as these.

1. _____

   _____

2. _____

   _____

When you have described two instructional activities, turn to page 166 and compare your suggestions with those listed there.

Error Pattern S-F-3 *(from Ann's paper on page 82)*

Did you find the pattern?

D.
$$5\frac{3}{8} = 5\frac{43}{8}$$
$$-2\frac{1}{2} = 2\frac{5}{8}$$
$$\overline{\qquad 3\frac{38}{8}}$$

E.
$$4\frac{1}{3} = 4\frac{13}{15}$$
$$-1\frac{4}{5} = 1\frac{9}{15}$$
$$\overline{\qquad 3\frac{4}{15}}$$

Ann has difficulty changing mixed numbers to equivalent mixed numbers that have a common denominator. She does determine a common denominator, but she computes each new numerator by multiplying the original denominator times the whole number and adding the original numerator. She merely copies the given whole number.

You probably recognize part of the procedure as the way to find the new numerator when changing a mixed number to a fraction, but doing this makes no sense when changing a mixed number to an equivalent mixed number with a specified denominator.

How would you help Ann? Describe two instructional activities you believe will help her subtract correctly when she encounters examples such as these.

1. _____

_____

2. _____

_____

If you have described two activities, turn to page 167 and compare what you have written with the suggestions recorded there.

Error Pattern M-F-1 *(from Dan's paper on page 83)*

Did you find Dan's error pattern?

E. $\dfrac{3}{4} \times \dfrac{2}{3} = 89$ F. $\dfrac{4}{9} \times \dfrac{2}{5} = 200$

Dan begins by multiplying the first numerator and the second denominator and recording the units digit of this product. If there is a tens digit, he remembers it to add later (as in multiplication of whole numbers). He then multiplies the first denominator and the second numerator, adds the number of tens remembered, and records this as the number of tens in the answer. The procedure involves a sort of cross multiplication and the multiply-then-add sequence from multiplication of whole numbers.

Dan uses this error pattern consistently. He has somehow learned to multiply fractions this way. How would you help him learn the correct procedure? Describe two instructional activities you believe would help.

1. _____

_____

2. _____

_____

After two activities have been described, turn to page 169 and compare what you have written with the suggestions listed there.

Error Pattern M-F-2 *(from Grace's paper on page 84)*

Did you identify the error pattern Grace is using?

D. $\dfrac{2}{3} \times \dfrac{3}{4} = \dfrac{2}{3} \times \dfrac{4}{3} = \dfrac{8}{9}$

E. $\dfrac{5}{7} \times \dfrac{3}{8} = \dfrac{5}{7} \times \dfrac{8}{3} = \dfrac{40}{21}$

Grace has learned to invert and multiply, and she is using this division procedure to multiply fractions.

How would you help Grace? Describe two instructional activities that would help her replace this error pattern with the correct computational procedure.

1. _____

_____

2. _____

_____

After you complete your two descriptions, turn to page 169 and compare what you have written with the suggestions recorded there.

Error Pattern M-F-3 *(from Lynn's paper on page 85)*

Did you find the procedure Lynn is using?

E. $\dfrac{3}{8} \times 4 = \dfrac{12}{32}$ F. $\dfrac{5}{6} \times 2 = \dfrac{10}{12}$

Lynn has learned that when you are multiplying and you have a fraction, you have to multiply *both* the numerator and the denominator of the fraction. Of course, when multiplying both terms by the same number she is actually multiplying the fraction by *one* rather than by the whole number in the example. It may be that she multiplied both numerator and denominator by the same number when she practiced changing a fraction to higher terms, and she continues to use this familiar pattern.

How would you help Lynn? Describe two activities that you believe would enable her to multiply correctly.

1. _____

_____

2. _____

_____

If you have described two such activities, turn to page 170 and compare your suggestions with the ideas listed there.

Error Pattern D-F-1 *(from Linda's paper on page 86)*

Did you find the error pattern Linda is using?

E.
$$\frac{4}{12} \div \frac{4}{4} = \frac{1}{3}$$

F.
$$\frac{13}{20} \div \frac{5}{6} = \frac{2}{3}$$

Linda divides the first numerator by the second numerator and records the result as the numerator for the answer. She then determines the denominator for the answer by dividing the first denominator by the second denominator. In both divisions she ignores remainders. Note that examples A, B, and E are correct. It may be that she learned her procedure while the class was working with such examples. The common denominator method of dividing fractions may be part of the background because her procedure is similar; however, she fails to change the fractions to equivalent fractions with the same denominator before dividing.

This is a tricky error pattern, producing both correct answers and absurd answers with zero numerators and denominators. Linda obviously needs help. How would you help her? Describe two instructional activities you think would enable her to replace her error pattern with a correct computational procedure.

1. _____

_____

2. _____

_____

When you have noted your descriptions, turn to page 171 and compare them with the suggestions listed there.

Error Pattern D-F-2 *(from Joyce's paper on page 87)*

Did you find the procedure Joyce is using?

$$\text{D.}\quad \frac{5}{8} \div \frac{2}{3} = \frac{8}{5} \times \frac{2}{3} = \frac{16}{15}$$

$$\text{E.}\quad \frac{1}{2} \div \frac{1}{4} = \frac{2}{1} \times \frac{1}{4} = \frac{2}{4}$$

Joyce knows to "invert and multiply," but she inverts the dividend (or product) instead of the divisor. It *does* make a difference!

Dividing fractions seems to involve such an arbitrary rule. How would *you* help Joyce? Describe two instructional activities you believe would help her divide correctly.

1. _____

_____

2. _____

_____

When you have described both activities, turn to page 172 and compare your ideas with the activities described there.

Error Pattern A-D-1 *(from Harold's paper on page 88)*

Did you find Harold's error pattern?

$$E. \quad \begin{array}{r} .3 \\ +.5 \\ \hline .8 \end{array} \qquad F. \quad \begin{array}{r} .7 \\ +.7 \\ \hline .14 \end{array}$$

     Harold seemingly adds these decimals as he would add whole numbers. The placement of the decimal point in the sum is a problem, but in every case he merely places the decimal point at the left of the sum. In example F he might explain that seven tenths plus seven tenths is 14 tenths.

     We ought to be able to help Harold with a problem of this sort. How would *you* help him? Describe at least two instructional activities you believe would enable Harold to add such examples correctly.

1. _____

   _____

2. _____

   _____

     After you have recorded both activities, turn to page 173 and compare your suggestions with the suggestions listed there.

Error Pattern S-D-1 *(from Les's paper on page 89)*

Did you determine the procedure Les is using?

$$D. \quad 60 - 1.35 = ? \qquad E. \quad 24.8 - 2.26 = ?$$

$$\begin{array}{r} 60 \\ - 1.35 \\ \hline 59.35 \end{array} \qquad \begin{array}{r} 24.8 \\ - 2.26 \\ \hline 22.66 \end{array}$$

When presented with "ragged" decimals such as these, Les simply brings down the extra digits at the right. (This procedure produced an acceptable result when he added similar decimals!) Apparently, when there are no ragged decimals he is able to subtract correctly.

What would you do to help Les? Describe two activities you believe would help him subtract correctly when confronted with ragged decimals.

1. _____

_____

2. _____

_____

When your activities have been described, turn to page 175 and compare your ideas with the suggestions listed there.

Error Pattern M-D-1 *(from Marsha's paper on page 90)*

If you used Marsha's error pattern, you completed examples E and F as they are shown.

E.
$$\begin{array}{r} 40.5 \\ \times\ \ .6 \\ \hline 24.30 \end{array}$$

F.
$$\begin{array}{r} 6.7 \\ \times\ \ 3 \\ \hline 2.01 \end{array}$$

In her answer, Marsha places the decimal point by counting over from the left instead of from the right in the product. She frequently gets the correct answer (as in examples A, B, and E), but much of the time her answer is not the correct product.

If you were Marsha's teacher, what corrective procedures might you follow? Describe two instructional activities that you think would help Marsha multiply decimals correctly.

1. _____

_____

2. _____

_____

When your responses are complete, turn to page 176 and see if your suggestions are among the alternatives described.

Error Pattern D-D-1 *(from Ted's paper on page 91)*

If you found Ted's error pattern, your results are the same as the erroneous computations shown.

D.
$$3\overline{)2.57} \quad .852$$

$$\begin{array}{r} 24 \\ \hline 17 \\ 15 \\ \hline 2 \end{array}$$

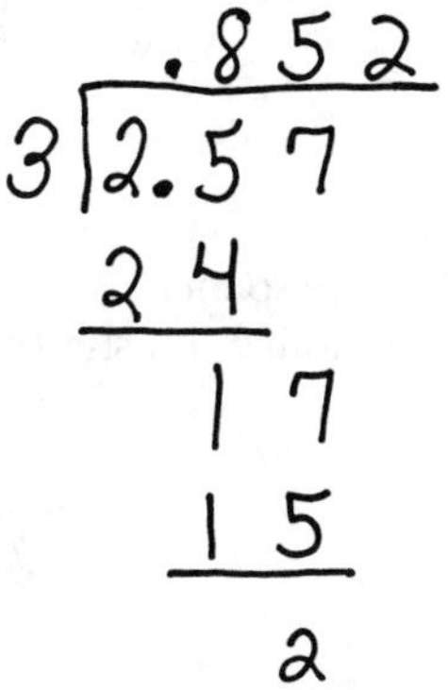

E.
$$.7\overline{)9.35} \quad 13.34$$

$$\begin{array}{r} 7 \\ \hline 23 \\ 21 \\ \hline 25 \\ 21 \\ \hline 4 \end{array}$$

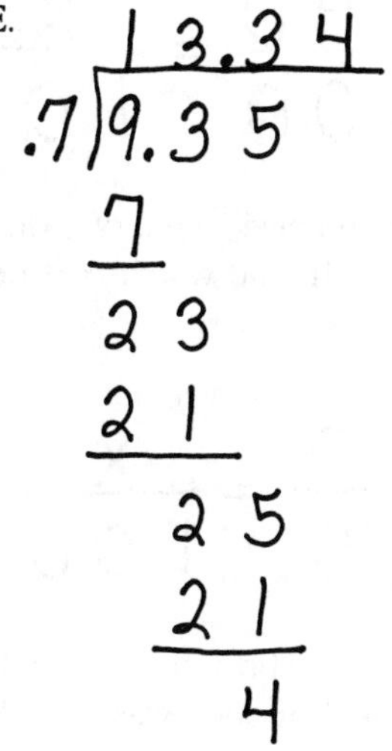

Ted misses examples because of the way he handles remainders. If division does not "come out even" when taken as far as digits given in the dividend, Ted writes the remainder as an extension of the quotient. He may believe this is the same as writing R2 after the quotient for a division problem with whole numbers. Some children, having studied division with decimals, thereafter use a procedure similar to Ted's when dividing whole numbers. For example, 400 divided by 7 is computed as 57.1.

How would you help Ted? Describe at least two instructional activities you believe would correct his error pattern.

1. _____

_____

2. _____

_____

After you have described at least two activities, turn to page 176 and compare your suggestions with the suggestions listed there.

Error Pattern P-P-1 *(from Sara's paper on page 92)*

If you used Sara's error pattern, you completed the three percent problems as shown.

D. Brad earned $400 during the summer, and saved $240 from his earnings. What percent of his earnings did he save?

$$\frac{240}{400} = \frac{X}{100}$$ Answer. __60%__

E. Barbara received a gift of money on her birthday. She spent 80% of the money on a watch. The watch cost her $20. How much money did she receive as a birthday gift?

$$\frac{20}{80} = \frac{X}{100}$$ Answer: __$25__

F. The taffy sale brought in a total of $750, but 78% of this was used for expenses. How much money was used for expenses?

$$\frac{78}{750} = \frac{X}{100}$$ Answer: __$10.40__

Sara successfully solved percent problems when the class first solved them, but as different types of problems were encountered she began to have difficulty. She *is* correctly solving the proportion she writes for the problem.

However, Sara uses a procedure that often does not accurately represent the ratios described in the problem. She is using the following proportion for every problem encountered:

$$\frac{\text{lesser number in the problem}}{\text{greater number in the problem}} = \frac{x}{100}$$

She appears to have abstracted the procedure from initial experiences with problems like A and D, with percents of less than 100. In each of the problems encountered, the first ratio was the lesser number over the greater number; this was always equal to $x/100$. Interestingly, the procedure also seems to work with problems of the type illustrated by B and E, even though different ratios are suggested; however, the procedure does not work with problems of the type illustrated by C and F.

How would you help Sara? Describe at least two instructional activities you believe would help her correctly solve percent problems.

1. _____

_____

2. _____

_____

After you have described at least two activities, turn to page 178 and compare your suggestions with those listed there.

Error Pattern P-P-2 (from Steve's paper on page 93)

If you found Steve's error pattern, your results are as follows.

D. What number is 80% of 54?

$$
\begin{array}{r}
5\,4 \\
\times\,.8\,0 \\
\hline
0\,0 \\
4\,3\,2 \\
\hline
4\,3.2\,0
\end{array}
$$

Answer: 43.2

E. Seventy is 14% of what number?

$$
\begin{array}{r}
7\,0 \\
\times\,.1\,4 \\
\hline
2\,8\,0 \\
7\,0 \\
\hline
9.8\,0
\end{array}
$$

Answer: 9.8

F. What percent of 125 is 25?

$$
\begin{array}{r}
1\,2\,5 \\
\times\ .2\,5 \\
\hline
6\,2\,5 \\
2\,5\,0 \\
\hline
3\,1.2\,5
\end{array}
$$

Answer: 31.25

Steve's solutions are correct when he is finding the percent of a specified number (problems A and D). However, his solutions are incorrect when the percent is known and he is to find a number (problems B and E) and when he is to find what percent one number is of a specified number (problems C and F).

Usually, the first percent problems a student encounters involve finding the percent of a number. Steve probably developed his procedure while solving such problems, but he is also using a somewhat updated version of it when he attempts to solve other types of percent problems. When the percent is given, he changes it to a decimal, then multiplies this number times the other number given. When the percent is not given, he treats the lesser of the two given numbers as if it were a decimal and proceeds similarly.

How would you help Steve? Describe at least two instructional activities you believe would help him correctly solve percent problems.

1. _____

   _____

2. _____

   _____

After you have written at least two descriptions, turn to page 180 and check your suggestions against those listed there.

Error Pattern S-M-1 *(from Margaret's paper on page 95)*

Using Margaret's error pattern, examples C and D would be completed as shown.

Margaret is regrouping in order to subtract just as she does when subtracting whole numbers expressed with base 10 numeration. She crosses out the left figure and writes one less above it, then places a one in front of the right figure. This technique produces a correct result when the relationship between the two measurement units is a base 10 relationship, but the results are incorrect whenever other relationships exist.

C.

$$\begin{array}{r} \overset{5}{6} \text{ yards, } 1 \text{ foot} \\ -2 \text{ yards, } 2 \text{ feet} \\ \hline 3 \text{ yards, } 9 \text{ feet} \end{array}$$

D.

$$\begin{array}{r} \overset{2}{3} \text{ quarts, } 1 \text{ cup} \\ -1 \text{ quart, } 3 \text{ cups} \\ \hline 1 \text{ quart, } 8 \text{ cups} \end{array}$$

How would *you* help Margaret? Describe at least two instructional activities you believe would make it possible for Margaret to subtract correctly in measurement situations.

1. _____

_____

2. _____

_____

After you have written at least two descriptions, turn to page 182 and see if any of your activities are among the suggestions listed there.

ENDNOTES

1. L. S. Cox, "Systematic Errors in the Four Vertical Algorithms in Normal and Handicapped Populations," *Journal for Research in Mathematics Education* 6, no. 4 (November 1975): 202–20.
2. Cox, *Ibid.*
3. Cox, *Ibid.*
4. Cox, *Ibid.*
5. Cox, *Ibid.*
6. Cox, *Ibid.*
7. Francis G. Lankford, Jr., *Some Computational Strategies of Seventh Grade Pupils,* U.S. Department of Health, Education, and Welfare, Office of Education. National Center for Educational Research and Development (Regional Research Program) and The Center for Advanced Study, The University of Virginia, October 1972, p. 30. (Project number 2-C-013. Grant number OEG-3-72-0035)

Helping Children
Correct Error Patterns
in Computation

Whenever an error pattern is identified in a child's written work, corrective instruction needs to be provided so he will be able to replace his erroneous procedure with a useful algorithm. Only then can he make adequate use of computational procedures as tools. When a child solves mathematical problems and explores topics of interest, skill in computation permits him to expand his horizons conceptually.

In Chapter 2 a list of guidelines for instruction was presented. The following chapter contains descriptions of instructional activities that may be useful in helping children correct specific patterns of errors. It is always important for the teacher to have in mind more than one instructional strategy; therefore, several activities are suggested for each error pattern. As you examine an error pattern, consider your own suggestions for providing instruction. Do you find your suggestions among those recorded on the succeeding pages?

Specific ideas for corrective instruction can also be found in many of the references listed after this chapter.

Error Pattern A-W-1 *(from pages 56 and 96)*

What instructional activities did you suggest to help correct the error pattern illustrated? See if your suggestions are among those listed.

E.
$$\begin{array}{r} 43 \\ +65 \\ \hline 108 \end{array}$$

F.
$$\begin{array}{r} 88 \\ +39 \\ \hline 1117 \end{array}$$

Note: Be sure to extend your diagnosis by interviewing the student. Let Mike "think out loud" for you. Unless you do this you will not even know if he is adding the ones or the tens first.

1. *Use bundles of ten and single sticks.* Show both addends, then "make a ten" as may have been done in past instruction. Emphasize that we always need to start with the single sticks. Apply a rule for exchanging or regrouping if it is possible. Then make ten bundles of ten, if possible, and apply the rule again. With paper and pencil, record what is done *step-by-step*.

2. *Provide the student with a set of numerals (0–9) and a frame for the answer.* Each box of the frame should be of a size that will enclose only one digit. Let the student use the cardboard or plastic numerals to record sums for problems. This activity should help the student remember to apply the rule for exchanging.

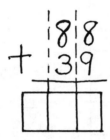

3. *Play chip-trading games.* To develop the idea for exchanging many for one, play games in which the values of chips are defined in terms of our numeration place-value pattern. However, it is easier to begin with bases less than ten. A child rolls a die and receives as many units as indicated on the die. He then exchanges for higher valued chips according to the rule of the game (five for one if base five, ten for one if base ten). Play proceeds similarly. The first child to get a specific chip of a high value wins. Such games are described in *Chip Trading Activities, Book I.*[1]

Error Pattern A-W-2 *(from pages 57 and 97)*

What instructional activities do you suggest to help Mary correct the error pattern illustrated? See if your suggestions are among those described.

E.
$$
\begin{array}{r}
254 \\
+\ 535 \\
\hline
789
\end{array}
$$

F.
$$
\begin{array}{r}
{\scriptstyle 3\,2} \\
618 \\
+\ 782 \\
\hline
1112
\end{array}
$$

1. *Estimate sums.* Even *before* computing, the sum can be estimated. For instance, in example F it can be determined in advance that the sum is more than 1300.

2. *Use a game board and a bank.*[2] Help children understand place values and begin computation with units by making the algorithm a record of moves in a game (or a problem to solve). If base ten blocks are used on a gameboard, the object of the game is to show the amount of wood on the board with as few pieces as possible. To compute example F, the child places 6 hundreds, 1 ten, and 8 unit blocks in the upper row. Then she sorts blocks for 782 in the second row. Beginning with the units (at the arrow), the child collects 10 units if she can

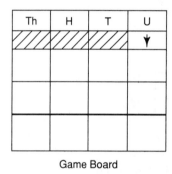

Game Board

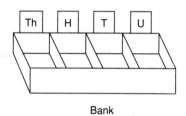

Bank

(for this is a base ten game) and moves all remaining units below the heavy line. If she has been able to collect 10 units, they are traded at the bank for 1 ten; the ten is then placed above the other tens in the shaded place. (At first many children want to verify the equivalence of what goes into the bank and what comes out by placing the 10 units in a row and matching them with 1 ten.) As the child continues, she collects 10 tens if she can, and then moves all remaining tens below the heavy line. If she has been able to collect 10 tens, they are traded at the bank for 1 hundred, and the hundred is placed above the other hundreds in the shaded place. Finally, the child collects 10 hundreds if she can and moves all remaining hundreds below the heavy line. If she has been able to collect 10 hundreds they are traded at the bank for 1 thousand, and the thousand is placed in the shaded area at the top of the column for thousands. It is not possible to collect 10 thousands, so the one remaining thousand is brought below the heavy line. As the child computes the sum on paper, she records the number of blocks in each region every time trading is completed. (Making the record can be likened to writing a story.) When the record is finished, the algorithm is completed.

Th	H	T	U
////////	////////	////////	↓
	□ □ □ □ □ □	▯	◻◻◻◻ ◻◻◻◻
	□ ◻□ □ □ ◻	▯▯▯▯ ▯▯▯▯	◻ ◻

Error Pattern A-W-3 *(from pages 58 and 98)*

You have written suggestions for helping Carol, who was using the error pattern illustrated. Are your suggestions among those listed?

F.
$$\begin{array}{r} 2\,6 \\ +\ \ 3 \\ \hline 1\,1 \end{array}$$

G.
$$\begin{array}{r} 6\,0 \\ +\,2\,4 \\ \hline 8\,4 \end{array}$$

H.
$$\begin{array}{r} 7\,4 \\ +\ \ 5 \\ \hline 1\,6 \end{array}$$

Note: An interview with the child may provide very helpful information. Is the child able to explain the examples that were worked correctly? Does the child identify tens and units and reason that units must be added to units and tens must be added to tens?

1. *Play "Pick-a-Number."* This game stresses the different values a digit may signify in various positions. Use cards with 0–9. Each player makes a form like this:

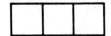

One player picks a card and reads the number, and each player writes that number in one of the spaces. Repeat until all blanks are full. The player showing the greatest number wins.

2. *Show each addend with base ten blocks.* After the child shows both addends, have her collect the units and record the total number of units. She can then collect the tens and record the total number of tens.

3. *Show addends with sticks or toothpicks.* Bundles of ten and single sticks (or toothpicks) can be used. Proceed as with base ten blocks.

4. *Draw a line to separate tens and units.* This procedure may help with the mechanics of notation if the child understands the need to add units to units and tens to tens.

$$
\begin{array}{c|c}
T & U \\
 & 3 \\
+\ 2 & 6 \\
\hline
2 & 9
\end{array}
\qquad
\begin{array}{c|c}
T & U \\
6 & 0 \\
+\ 2 & 4 \\
\hline
8 & 4
\end{array}
\qquad
\begin{array}{c|c}
T & U \\
7 & 4 \\
+\ & 5 \\
\hline
7 & 9
\end{array}
$$

Error Pattern A-W-4 *(from pages 59 and 99)*

You have suggested instructional activities for helping Dorothy or any child using this error pattern. Are your suggestions among those listed?

E.
$$
\begin{array}{r}
{}^{1}4\ 6 \\
+\quad 8 \\
\hline
1\ 3\ 4
\end{array}
$$

F.
$$
\begin{array}{r}
{}^{1}9\ 8 \\
+\quad 3 \\
\hline
1\ 3\ 1
\end{array}
$$

Note: The following suggestions assume that the child is *not* confusing these higher decade situations with multiplication.

1. *Explain the process naming units and tens.* Have the child explain the addition to you in terms of units (or ones) and tens. If her understanding of place value is adequate, this procedure may be sufficient to clear up the difficulty. It may be necessary to have the child use base ten blocks or a place-value chart.

2. *Label units and tens.* Have the child label each column. The use of squared paper may also help if only one digit is written in each square.

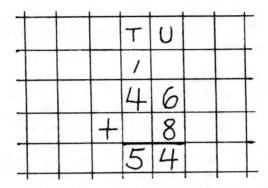

3. *Make higher decade sequences.* Help the child discover the pattern illustrated, then have her complete similar sequences. She may want to make up a few patterns all on her own.

$$\begin{array}{r} 6 \\ + 8 \\ \hline 14 \end{array} \qquad \begin{array}{r} 16 \\ + 8 \\ \hline 24 \end{array} \qquad \begin{array}{r} 26 \\ + 8 \\ \hline 34 \end{array} \qquad \begin{array}{r} 36 \\ + 8 \\ \hline 44 \end{array} \ \cdots$$

Have you examined the papers in Appendix A that involve adding whole numbers? See if you can find the error patterns.

Error Pattern S-W-1 *(from pages 60 and 100)*

You have described instructional activities to help correct the error pattern illustrated. Are your activities among those described?

$$\text{E.} \quad \begin{array}{r} 458 \\ -372 \\ \hline 126 \end{array} \qquad\qquad \text{F.} \quad \begin{array}{r} 241 \\ -96 \\ \hline 255 \end{array}$$

Note: Be sure to extend your diagnosis by interviewing the child and letting her think out loud as she works similar examples. Does she use the erroneous procedure only with greater numbers? Does the child, on her own initiative, question the reasonableness of her answers? (In example F, the result is greater than the sum.)

1. *Use bundles of 100, bundles of ten and single sticks.* Let the student show the "number altogether," that is, the minuend or sum shown by the upper numeral. Pose the problem of removing the number of sticks shown by the lower numeral. Any verbal problems presented in this context should describe take away rather than comparison situations. Trading or exchanging as needed could be done at a trading post or a bank. (If more than one child is having this difficulty, pose the problem to a pair of children who will work on the task together.) Eventually, guidance should be provided to help the student remove units first, then tens, etc.

2. *Use base ten blocks.* Proceed as in activity 1.

3. *Use a place-value chart.* Proceed similarly.

Error Pattern S-W-2 *(from pages 61 and 101)*

You have described two instructional procedures for helping a child using this error pattern. Are the activities you suggested for George similar to any of the activities described?

D.
$$2 \overset{6}{\cancel{7}} \overset{1}{3} \\ -\quad 3\ 8 \\ \hline 2\ 3\ 5$$

E.
$$2 \overset{7}{\cancel{8}} \overset{1}{5} \\ -\quad 6\ 3 \\ \hline 2\ 1\ 1\ 2$$

Note: Helpful instruction will emphasize (1) the ability to distinguish between subtraction problems requiring regrouping in order to use basic subtraction facts and subtraction problems not requiring regrouping, and (2) mechanics of notation.

1. *Use a physical representation for the minuend (sum).* If the minuend of example E is represented physically (with base blocks or bundles of sticks), questions can be posed such as "Can I take away 3 units *without* trading?" or "When do I need to trade and when is it not necessary for me to trade?"

2. *Replace computation with* yes *or* no. Focus on the critical skill of distinguishing by presenting a row of subtraction problems for which the differences are *not* to be computed. Have the child simply write yes or no for each example to indicate the decision whether regrouping is or is not needed. If this is difficult, physical materials should be available for the child to use. (See activity 1.)

3. *Estimate differences.* Before each problem is solved, have the student estimate the answer. Encourage statements like "more than 300" or "less than 500" rather than exact answers.

Error Pattern S-W-3 *(from pages 62 and 102)*

You have described two instructional activities you think would help children like Donna with this zero difficulty. Are any of your suggestions among those listed?

E.
$$
\begin{array}{r}
4\ 4\ 6 \\
-\ 3\ 0\ 2 \\
\hline
1\ 0\ 4
\end{array}
$$

F.
$$
\begin{array}{r}
7\ 6\ 0 \\
-\ 2\ 3\ 0 \\
\hline
5\ 3\ 0
\end{array}
$$

1. *Use sets and record number sentences.* In a demonstration with simple subtraction facts show the sum (the minuend) with a set of objects. Next, remove a subset of as many objects as the known addend. Finally, record the missing addend as the number of objects remaining. After demonstrating the process, let the child having difficulty remove a subset from a set of nine objects or less while you record the number sentence as a record of what is done. Encourage the child to remove the empty set. Then reverse the process and you demonstrate while the child records with a number sentence. Include many examples of removing the empty set.

2. *Use base blocks or bundled sticks to picture the computation.* Show the sum (the minuend) of a given subtraction problem. Sit beside the child and arrange

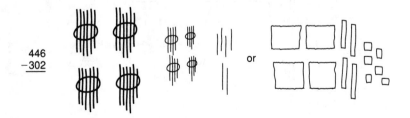

the materials so that units are to the right and hundreds to the left as in the algorithm. For example E, beginning with the units, have the child remove the number (of sticks or blocks) shown by the given addend. After the child removes a subset of 2 units, you record the fact that 4 units remain. After the child removes an empty set of tens, you record the fact that 4 tens remain, etc. For another example, you remove the subsets while *the child* records the number remaining each time.

3. *Try a sorting game.* If the child has been introduced to the multiplication facts for zero, there may be confusion between the zero property for multiplication and the zero properties for other operations. For the zero facts of arithmetic, prepare cards showing open number sentences similar to the ones shown.

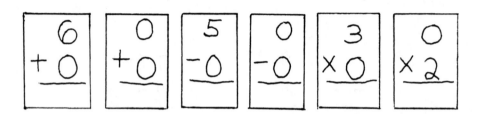

(Vertical notation is suggested in this case because it appears in the algorithm. It may be appropriate to include division number sentences as well, *e.g.,* $0 \div 3 = ?$) As an individual activity or as a game for two, the cards can be sorted into two sets—those with zero for the answer and those that do not have zero for an answer.

4. *Use a calculator.* Have the child try many examples of adding a zero, subtracting a zero, and multiplying by a zero. Then ask the child to state a generalization or rule for each operation. Also have the child compare the rules to see how they are alike or different.

Error Pattern S-W-4 (from pages 63 and 103)

Do you find, among the suggestions listed, your suggestions for helping Barbara, who has the difficulty illustrated?

E.
$$\begin{array}{r} \overset{3}{\cancel{4}}\,\overset{1}{3}\,6 \\ -\,1\,7\,2 \\ \hline 2\,6\,4 \end{array}$$

F.
$$\begin{array}{r} \overset{4}{\cancel{6}}\,\overset{1}{2}\,\overset{1}{5} \\ -\,3\,4\,8 \\ \hline 1\,8\,7 \end{array}$$

Note: Asking a child to check his computation may accomplish little in this situation. "Adding up" may only confirm that individual subtraction facts have been correctly completed. (In example F, $7 + 8 = 15$, $8 + 4 = 12$, and $1 + 3 = 4$.) The following activities are suggested to help the child keep in mind the *total quantity* from which a lesser number is being subtracted.

1. *Use base ten blocks to show the sum (minuend).* Pointing to the known addend (subtrahend) ask, "Do we need to trade so we can remove this many? What trading must we do?" As appropriate, trade a ten for ones and *immediately* record the action; then trade a hundred for tens and record that action. Stress the need to proceed step by step. Have the child trade and remove blocks while you record the action, then reverse the process. While you trade and remove blocks (thinking aloud as you do) have the child make the record.

2. *Use a place-value chart.* Proceed in a manner similar to that suggested for base ten blocks.

3. *Use bundles of 100, bundles of ten and single sticks.* Proceed as in activity 1.

4. *Use expanded notation.* Have the child rename the sum to a name which makes possible the use of basic subtraction facts. In this example, $600 + 20 + 5$ was renamed as $600 + \underline{10 + 15}$, then $\underline{600 + 10} + 15$ was renamed as $\underline{500 + 110} + 15$. Base ten blocks can be used to verify such equivalences.

$$\begin{array}{r}
& 500 & \overset{110}{\cancel{20}} & 15 \\
6\,2\,5 = & \cancel{600}\; + & \cancel{20}\; + & \cancel{5} \\
-3\,4\,8 = & 300\; + & 40\; + & 8 \\
\hline
& 200\; + & 70\; + & 7 = 277
\end{array}$$

Error Pattern S-W-5 (from pages 64 and 104)

Are your suggestions for helping Sam among the suggestions listed? He has the illustrated difficulties.

E.
$$\begin{array}{r} \overset{4}{\cancel{3}}\,\overset{7}{\cancel{8}}\,5 \\ -\ 3\ 2\ 2 \\ \hline 1\ 5\ 3 \end{array}$$

F.
$$\begin{array}{r} \overset{7}{\cancel{8}}\,\overset{3}{\cancel{4}}\,0 \\ -\ 6\ 2\ 6 \\ \hline 1\ 1\ 0 \end{array}$$

1. *Use base blocks or bundled sticks to picture the sum (minuend).* Specific procedures are outlined with Error Pattern S-W-4.

2. *Estimate before computing.* Have the child estimate his answer. A number line showing at least hundreds and tens may be helpful for this purpose. Ask, "Will the answer be more than a hundred? . . . less than a hundred?"

3. *Use a learning center for renaming.* For Sam and others with similar difficulties, a simple learning center could be set up to help them rename a minuend and select the most useful name for that number in a specific subtraction problem. One possibility is to have a sorting task in which the child decides which cards show another name for a given number and which show an entirely different number. A second task would be to consider all the different names for the given number and decide which of the names would be most useful for computing subtraction examples that have the given number as the minuend. Ask, "Which name will let us use the subtraction facts we know?"

4. *Use a game board and a bank.* Make the subtraction algorithm a record of moves in a game-like activity that can be presented as a problem to solve: the student is to show the amount of wood that will remain. Use the game board pictured and a bank with Error Pattern A-W-2 but have the child picture *only the minuend* with base blocks or similar materials. The subtrahend should be shown with numeral cards to indicate how much wood is to be removed from the set of wood used for the minuend. Begin with units and trade one ten for ten units if necessary; then place on the numeral card (in the subtrahend) the number of units indicated. Units that remain should be brought below the heavy line. For tens, repeat the process, trading one hundred for ten tens, if needed, in order to have enough tens pieces to go on the numeral card for tens. Then, bring any remaining tens below the line and continue the procedure for hundreds. Make a step-by-step record of the child's moves during the activity by recording each move in the evolving written algorithm. Later, let the child record your moves or those of another child.

Th	H	T	U
/////	/////	/////	↓
	□ □	□ □ □ □	□ □ □
	1	8	5

Game Board Used for Subtraction

Have you examined the papers in Appendix A that involve subtracting whole numbers? See if you can find the error patterns.

Error Pattern M-W-1 *(from pages 65 and 104)*

How would you help a student such as Bob correct the error pattern illustrated? Are the activities you described similar to any of those listed?

D.
$$\overset{4}{9}\ 8$$
$$\underline{5\ 6}$$
$$5\ 8\ 8$$
$$\underline{4\ 9\ 0}$$
$$5\ 4\ 8\ 8$$

E.
$$\overset{3}{8}\ 6$$
$$\underline{4\ 5}$$
$$4\ 3\ 0$$
$$\underline{3\ 5\ 4}$$
$$3\ 9\ 7\ 0$$

Note: The following activities emphasize place value, the distributive property, and proper mechanics of notation.

1. *Use more partial products and no crutch.* The algorithm that follows can be related to an array partitioned twice. When the student is able to use this algorithm with ease, let him try to combine the first two partial products (and also the last two) by *remembering* the number of tens (and the number of hundreds). Do not encourage the use of a crutch in this situation.

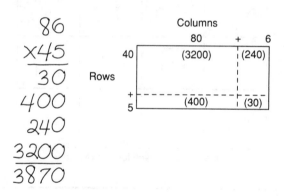

2. *Make two problems.* Have the student multiply by ones and then by tens in two separate problems. As the student computes the product in this way, encourage him to try remembering his crutch number (rather than writing it) "because such crutches are sometimes confusing in multiplication and division examples." When he can compute easily without recording the crutch, convert to a more standard algorithm by placing both products within *one* example.

3. *Record tens within partial products.* Instead of writing "crutch" numerals above the example, use lightly written, half-sized numerals within each of the partial products to record the number of tens to be remembered.

$$\begin{array}{r} 86 \\ \times\ 45 \\ \hline \end{array} \rightarrow \begin{array}{r} 86 \\ \times\ 45 \\ \hline {\scriptstyle 3}\ \ 0 \end{array} \rightarrow \begin{array}{r} 86 \\ \times\ 45 \\ \hline 4\overset{3}{3}0 \end{array}$$

$$\begin{array}{r} 86 \\ \times\ 45 \\ \hline 4\overset{3}{3}0 \\ {\scriptstyle 2}\ \ 4 \end{array} \rightarrow \begin{array}{r} 86 \\ \times\ 45 \\ \hline 4\overset{3}{3}0 \\ 3\overset{2}{4}4 \end{array} \rightarrow \begin{array}{r} 86 \\ \times\ 45 \\ \hline 4\overset{3}{3}0 \\ 3\overset{2}{4}4 \\ \hline 3870 \end{array}$$

4. *Apply the commutative and associative principles.* This technique should be especially helpful if the student has difficulty in processing open number sentences like $5 \times 80 = ?$; $40 \times 6 = ?$; and $40 \times 80 = ?$ These number sentences are parts of example E and suggest a prerequisite skill for such examples; namely, application of the commutative and associative principles where one of the factors is a multiple of a power of ten. The error pattern may result, in part, from thinking of all digits as ones and the inability to think of tens, hundreds, etc., when using basic multiplication facts. After the associative principle is introduced with one-digit factors (perhaps with the aid of a three-dimensional arrangement of cubic units) let the student think through examples such as:

$$\begin{aligned} 40 \times 6 &= (4 \times 10) \times 6 \\ &= (10 \times 4) \times 6 \\ &= 10 \times (4 \times 6) \\ &= 10 \times 24 \\ &= 240 \end{aligned}$$

When the student generalizes this procedure, he will be able to compute the product of a one-digit number and a multiple of a power of ten *in one step.*

Cautions

In general, written crutches are to be encouraged if they are useful and help the child understand what he is doing. However, they can be confusing when multiplying a two-digit multiplier. Also, questions such as 5 × 86 = ? occur within division algorithms where use of a written crutch is impractical. For these reasons, students should be encouraged to remember crutch numbers in multiplication.

There is a very real danger in proceeding to a standard algorithm too quickly. Often, a new, more efficient procedure is best introduced as a shortcut of an already understood algorithm. Whenever zeros help a student think in terms of place value, do not insist that the units zero in the second and succeeding partial products be dropped. (Pencil lead is not that expensive!)

Error Pattern M-W-2 *(from pages 66 and 105)*

In this illustrated error pattern, the child forgets to add the number of tens recorded as a crutch. Which of the instructional activities you suggested as help for this child are similar to activities described?

E.
$$
\begin{array}{r}
\overset{1}{3}5 \\
\times\ \ 3 \\
\hline
95
\end{array}
$$

F.
$$
\begin{array}{r}
\overset{3}{2}8 \\
\times\ \ 4 \\
\hline
82
\end{array}
$$

1. *Use partial products.* Introduce the following algorithm, possibly as a record of multiplying with parts of a partitioned array. If the child has previously used this algorithm and he uses it successfully, it may still be wise to return to the longer procedure so that instruction can proceed from a position of strength. Before returning to the standard algorithm, it may be helpful to have the child circle the number to be remembered.

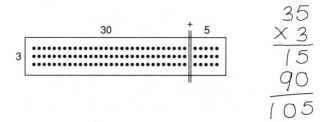

2. *Record a reminder below the bar.* Instead of a small numeral written above the multiplicand, introduce the idea that a reminder can be recorded as follows:

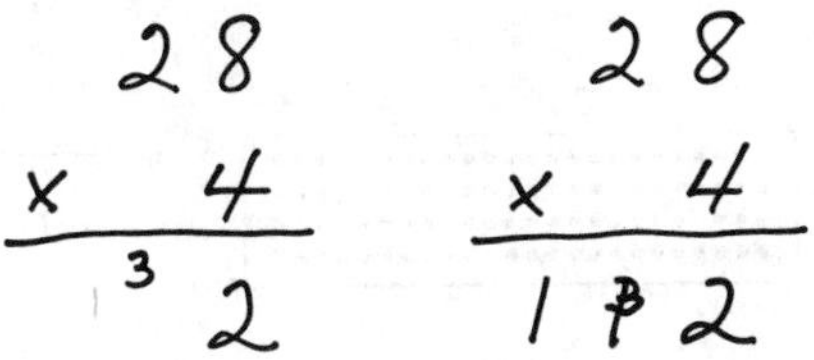

$$
\begin{array}{r}
28 \\
\times \quad 4 \\
\hline
^{3} \quad 2 \\
\end{array}
\qquad
\begin{array}{r}
28 \\
\times \quad 4 \\
\hline
1\,\overset{3}{?}\,2 \\
\end{array}
$$

This procedure is a convenient bridge between the algorithm using partial products and the standard algorithm.

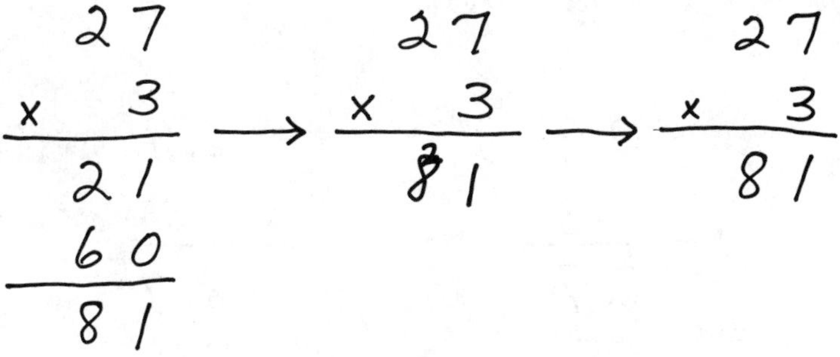

$$
\begin{array}{r}
27 \\
\times \quad 3 \\
\hline
21 \\
60 \\
\hline
81 \\
\end{array}
\longrightarrow
\begin{array}{r}
27 \\
\times \quad 3 \\
\hline
\overset{2}{8}\,1 \\
\end{array}
\longrightarrow
\begin{array}{r}
27 \\
\times \quad 3 \\
\hline
81 \\
\end{array}
$$

Error Pattern M-W-3 *(from pages 67 and 106)*

You have described two instructional activities for helping Joe and other students who have adopted the error pattern illustrated. Are your suggestions included in the activities described?

$$
\text{D.} \quad
\begin{array}{r}
\overset{4}{6}8 \\
\times \quad 5 \\
\hline
500 \\
\end{array}
\qquad\qquad
\text{E.} \quad
\begin{array}{r}
\overset{2}{2}9 \\
\times \quad 3 \\
\hline
127 \\
\end{array}
$$

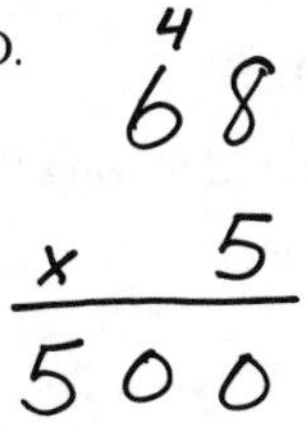

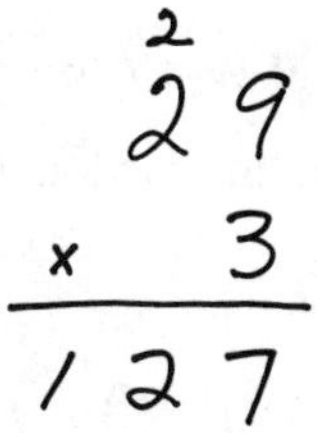

1. *Use partial products.* Such an algorithm is easily developed as a step-by-step record of what is done when an array is partitioned. Help the child determine the order in which the multiplication and addition occur; lead him to generalize

and state the sequence. Some children may find it helpful to whisper their thinking out loud as they work.

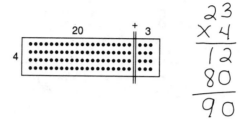

$$\begin{array}{r} 23 \\ \times\ 4 \\ \hline 12 \\ 80 \\ \hline 90 \end{array}$$

2. *Write the crutch below the bar.* Instead of the child writing a reminder in the conventional way, have him make a small numeral below the bar to remind him to add *just before* recording a product.

$$\begin{array}{r} 29 \\ \times\quad 3 \\ \hline {}^{2}\ 7 \end{array} \qquad\qquad \begin{array}{r} 29 \\ \times\quad 3 \\ \hline {}^{2}8\ 7 \end{array}$$

3. *Practice examples of the form* $(a \times b) + c.$ Sometimes a child appears to understand the algorithm and can verbalize the proper procedure correctly, but while using the algorithm he adopts careless procedures. Such a child may be helped by practicing examples such as $(4 \times 2) + 1 = ?$ and $(3 \times 6) + 2 = ?$, thereby reinforcing the proper sequence. The child should be helped to relate this kind of practice to his difficulty in the multiplication algorithm.

Error Pattern M-W-4 *(from pages 68 and 107)*

Are either of the instructional activities you suggested among those listed?

E.
$$\begin{array}{r} 621 \\ \times\quad 23 \\ \hline 1243 \end{array}$$

F.
$$\begin{array}{r} 5\overset{2}{1}7 \\ \times\ 463 \\ \hline 2081 \end{array}$$

1. *Use the distributive property.* Have the child rewrite each problem as two problems. Later relate each partial product to the partial products in the conventional algorithm.

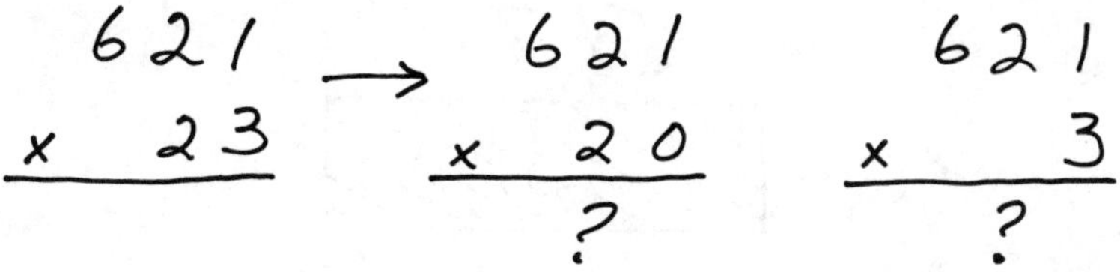

$$
\begin{array}{r} 621 \\ \times\quad 23 \\ \hline \end{array}
\;\longrightarrow\;
\begin{array}{r} 621 \\ \times\quad 20 \\ \hline ? \end{array}
\qquad
\begin{array}{r} 621 \\ \times\quad\ \ 3 \\ \hline ? \end{array}
$$

If the child does not understand why the sum of the two multiplication problems is the same number as the product in the original problem, partition an array and label the parts. If the child does not understand the concept of an array, begin with small numbers.

$$2 \times 3$$

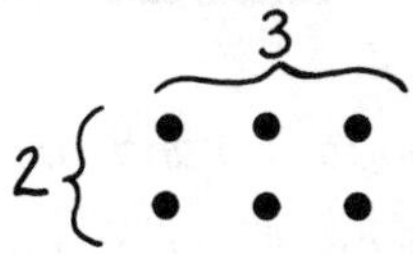

Then use rectangles to represent arrays with greater numbers.

$$23 \times 621$$

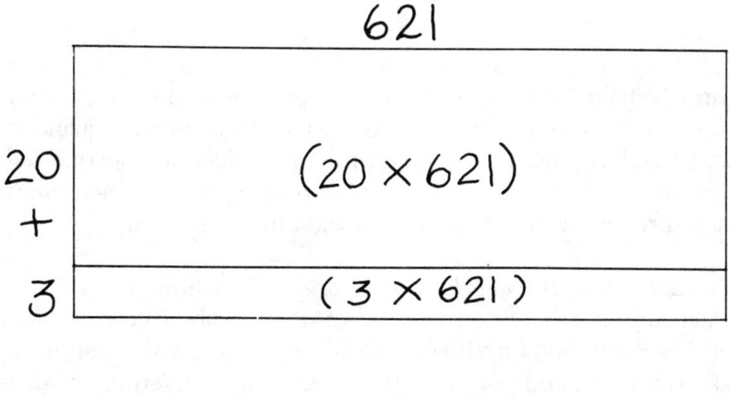

$$
\begin{array}{c|c}
 & 621 \\
\hline
20 & (20 \times 621) \\
+ & \\
3 & (3 \times 621) \\
\end{array}
$$

2. *Use a paper mask.* Cover the multiplier so only one digit will show at a time. After multiplication by the units digit is completed, the mask can be moved to the left so that only the tens digit is visible. Later, the hundreds digit can be

highlighted. With each digit, emphasize the need to do a complete multiplication problem. Also stress proper placement of each partial product.

3. *Use a calculator.* Use a calculator to compute each partial product. Make sure the correct values are multiplied: for 23 × 621, multiply 3 × 621 then 20 × 621.

> Have you examined the papers in Appendix A that involve multiplying whole numbers? See if you can find the error patterns.

Error Pattern D-W-1 *(from pages 69 and 108)*

How might you help Jim correct the erroneous procedure in the illustration? Are the instructional activities you described similar to any of the activities listed?

D.
$$\begin{array}{r} 2\ 1\ 3 \\ 3\overline{)6\ 3\ 9} \end{array}$$

E.
$$\begin{array}{r} 1\ 4\ 2 \\ 4\overline{)5\ 1\ 8} \end{array}$$

Note: To help the student who has adopted such an error pattern, activities need to emphasize place value in the dividend and the total quantity of the dividend. Procedures that can be understood in relation to concrete referents are needed instead of an assortment of rules to be applied in a mechanical way. In essence, a reintroduction to a division algorithm is needed.

1. *Use manipulatives to redevelop the algorithm.* Teach the child a computational procedure as a step-by-step record of activity with objects. For the problem 54 ÷ 3 = ?, the child can show the dividend as 5 tens and 4 units using base ten blocks, Cuisenaire rods, or single sticks and bundles. Interpret the divisor as the number of equivalent sets to be formed. For example, in 54 ÷ 3 = ?, distribute the objects equally among three sets. You may want to designate a collecting space for each of the three sets: sheets of paper, box lids, or similar places to put objects.

$$\begin{array}{r} 1 \\ 3\overline{)54} \\ \underline{3} \\ 2 \end{array}$$

$$\begin{array}{r} 1 \\ 3\overline{)54} \\ \underline{3} \\ 24 \end{array}$$

$$\begin{array}{r} 18 \\ 3\overline{)54} \\ \underline{3} \\ 24 \\ \underline{24} \end{array}$$

Step A   Step B   Step C

To begin, 1 ten is placed in each of the three sets. Then a record is made to show this has been done. The record should also show that a total of 3 tens has been removed from the dividend (Step A). As the two remaining tens cannot be distributed among three sets, they are traded for 20 units; the other four units are joined with them. This is shown by "bringing down" the "4" (Step B). The 24 units are then distributed equally among the three sets and the record is completed (Step C).

2. *Estimate quotient figures.* Use open number sentences such as $3 \times ? \leq 65$ with the rule that the number to be found is the largest multiple of a power of 10 that will make the number sentence true.

3. *Focus on skill in multiplying multiples of powers of ten by a single digit.* This skill, used in the above activities, may need to be developed independently of a division algorithm. Patterns can be observed from such data as the following display.

$$2 \times 3 = 6 \qquad\qquad 2 \times 3 = 6$$
$$2 \times 30 = 60 \qquad\qquad 20 \times 3 = 60$$
$$2 \times 300 = 600 \qquad 200 \times 3 = 600$$

Have the child state the pattern orally. Here is a possible conversation.

CHILD: You find the numbers [digits] for the multiplication fact, then you count the zeros and write the same number of zeros. Then you are done.

TEACHER: Yes, you multiply the single digit times the first digit in the other number and write that product, then count the zeros and affix that many zeros to your product. That's a great rule, but does it always work? Try your rule on these examples, then use your calculator to see if the rule always works.

It is possible to present a more detailed explanation related to a $2 \times 3 \times 10$ rectangular prism of unit cubes or to an application of mathematical principles.

$$2 \times 30 = 2 \times (3 \times 10)$$
$$= (2 \times 3) \times 10$$
$$= 6 \times 10$$
$$= 60$$

Error Pattern D-W-2 (from pages 70 and 108)

Illustrations of Gail's error pattern in division of whole numbers are shown here. Are the instructional activities you suggested to help this child among those described?

E.
```
      5 2
   3)7 5
     6 0
     ‾‾‾
     1 5
     1 5
```

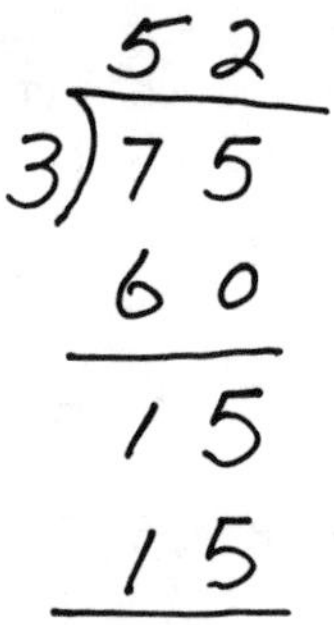

F.
```
        6 8
   6)5 1 6
     4 8 0
     ‾‾‾‾‾
       3 6
       3 6
```

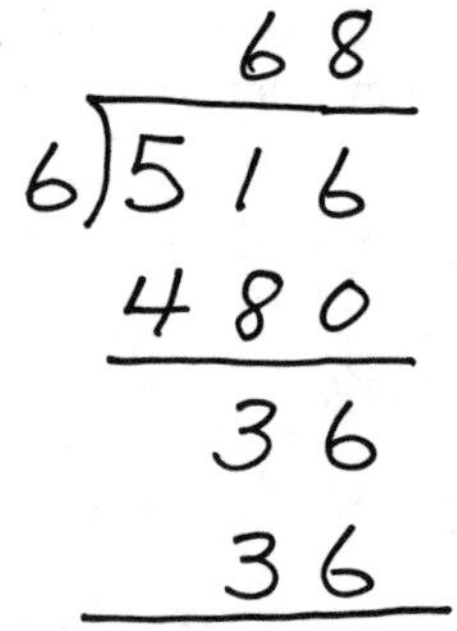

1. *Emphasize place value in estimating quotient figures.* Have the child use open number sentences such as $3 \times ? \le 70$ and $3 \times ? \le 15$ while thinking through example E. Number sentences should be completed using the following rule: When dividing 7 tens, the missing number is the largest multiple of ten that will make the number sentence true. For $3 \times ? \le 70$, $? = 20$. Similarly, when dividing 15 ones the missing number is the largest multiple of one that will make the number sentence true. For larger numbers, similar rules apply.

2. *Use a different algorithm.* At least temporarily, choose an algorithm that will show the value of each partial quotient. In each of the algorithms shown, the 8 in the quotient of example F is shown as 80, thereby emphasizing proper placement of quotient figures. After the student is able to use such algorithms, a transition to the standard computational procedure can be made, if desired, by recording quotients differently. In example F, the first quotient figure would be recorded as 8 in the tens place instead of as 80.

A.

$$
\begin{array}{r}
6\overline{)516} \\
480 = 80 \times 6 \\
36 \\
\underline{36 = 6 \times 6} \\
86 \times 6
\end{array}
$$

B.

$$
\begin{array}{r|l}
6\overline{)516} & \\
480 & 80 \\
\hline
36 & \\
\underline{36} & 6 \\
\hline
& 86
\end{array}
$$

C.

$$
\begin{array}{r}
86 \\
6 \\
80 \\
6\overline{)516} \\
480 \\
36 \\
\underline{36}
\end{array}
$$

3. *Develop skill in multiplying numbers of powers of 10.* This skill is necessary for rational use of any of the division algorithms illustrated. Exercises can be written to facilitate observations of patterns by the student.

$$
\begin{aligned}
6 \times 4 &= 24 \\
6 \times 40 &= 240 \\
6 \times 400 &= 2400
\end{aligned}
$$

Or, a more detailed explanation can be developed.

$$
\begin{aligned}
6 \times 400 &= 6 \times (4 \times 100) \\
&= (6 \times 4) \times 100 \\
&= 24 \times 100 \\
&= 2400
\end{aligned}
$$

4. *Estimate the quotient before computing.* Frequently, quotients resulting from the erroneous algorithm are quite unreasonable. If intelligent estimating is

followed by computing, and the estimate and the quotient are then compared, the student may rethink her computational procedure.

Error Pattern D-W-3 *(from pages 71 and 109)*

You have suggested activities for helping John, who was using the error pattern illustrated. Are your suggestions among those listed?

E.

$$
\begin{array}{r}
3\ 2\ r3 \\
9\overline{)2\ 7\ 2\ 1} \\
2\ 7 \\
\hline
2\ 1 \\
1\ 8 \\
\hline
3
\end{array}
$$

F.

$$
\begin{array}{r}
7\ 8\ r2 \\
6\overline{)4\ 2\ 5\ 0} \\
4\ 2 \\
\hline
5\ 0 \\
4\ 8 \\
\hline
2
\end{array}
$$

1. *Use lined paper turned 90°.* It may be that having the child use vertically lined paper (or cross-sectioned paper) will clear up the problem. The omission of one digit becomes very obvious when such forms are used for practice.

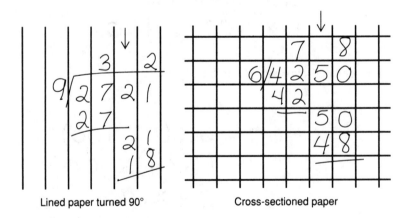

Lined paper turned 90°   Cross-sectioned paper

2. *Use the pyramid algorithm.* At least temporarily, use an algorithm that emphasizes place value. If the pyramid algorithm has been learned by the child earlier in the instructional program, ask him to solve some of the troublesome

examples using it to see if he can figure out why he is having difficulty now. If the pyramid algorithm is new to the child, he may enjoy trying a new procedure that is a bit easier to understand.

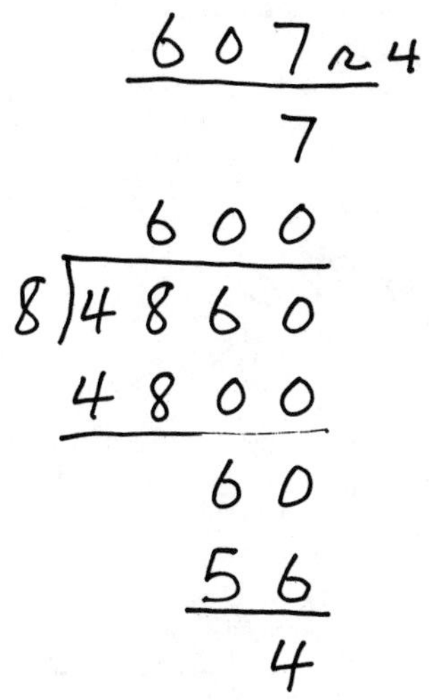

$$
\begin{array}{r}
6\ 0\ 7\ r\ 4 \\
7 \\
6\ 0\ 0 \\
8\overline{)4\ 8\ 6\ 0} \\
4\ 8\ 0\ 0 \\
\hline
6\ 0 \\
5\ 6 \\
\hline
4
\end{array}
$$

3. *Estimate the quotient before beginning computation.* The practice of recording an estimated quotient before computing may be sufficient to overcome the problem, especially if the error is not present in every such example and careless writing of quotient figures is a major cause of the difficulty.

Error Pattern D-W-4 *(from pages 72 and 110)*

Are your suggestions for helping a child like Anita with the error pattern shown, among those suggestions listed?

E.
$$
\begin{array}{r}
8\ 3\ 0 \\
6\overline{)4\ 8\ 1\ 8} \\
4\ 8\ 0\ 0 \\
\hline
1\ 8 \\
1\ 8 \\
\hline
\end{array}
$$

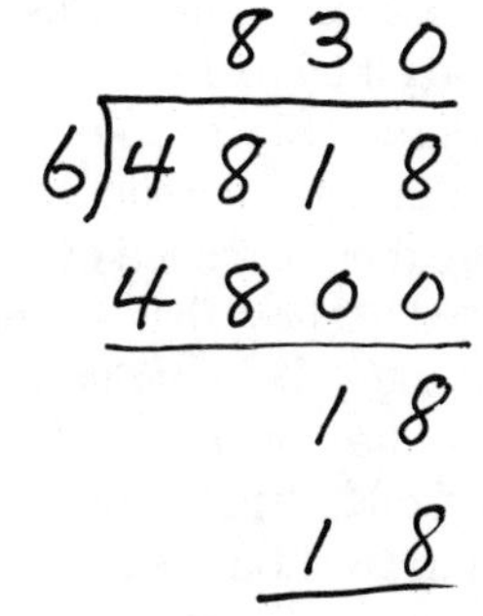

F.
$$
\begin{array}{r}
5\ 3\ 0\ r\ 4 \\
7\overline{)3\ 5\ 2\ 5} \\
3\ 5\ 0\ 0 \\
\hline
2\ 5 \\
2\ 1 \\
\hline
4
\end{array}
$$

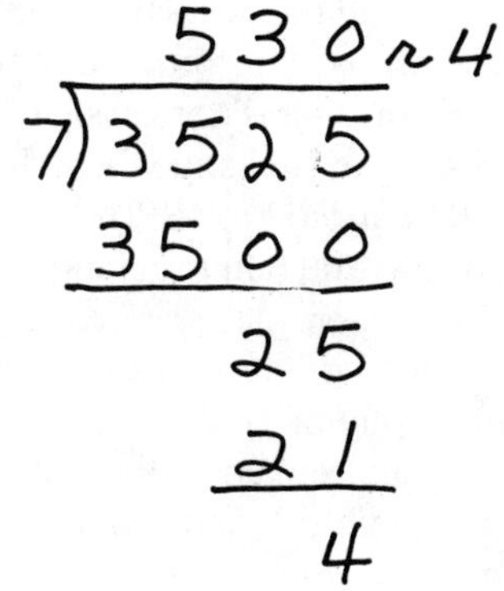

1. *Develop skill in multiplying multiples of powers of ten by a single-digit number.* Patterns such as this one can be observed. As the pattern is generalized and skill in such multiplication is developed, help the child see specific points within the division algorithm where this skill is applied.

$$3 \times 6 = 18 \qquad\qquad 7 \times 3 = 21$$
$$30 \times 6 = 180 \qquad 7 \times 30 = 210$$
$$300 \times 6 = 1800 \quad 7 \times 300 = 2100$$

2. *Use the pyramid algorithm.* Using an algorithm with partial quotients may adequately demonstrate the need for a zero in the tens place of the quotient.

$$
\begin{array}{r}
8\ 0\ 3 \\
\hline
3 \\
8\ 0\ 0 \\
6\,\overline{)4\ 8\ 1\ 8} \\
4\ 8\ 0\ 0 \\
\hline
1\ 8 \\
\hline
1\ 8
\end{array}
$$

3. *Use base ten blocks.* In the example 4818 ÷ 6 = ?, have the child show 4818 with 4 thousand-blocks, 8 hundred-blocks, 1 ten-block, and 8 unit-blocks. Write the problem, and interpret the problem as partitioning the blocks into 6 sets of equal number. As it is not possible to parcel out 4 thousand-blocks among 6 sets, it is necessary to exchange the 4 thousand-blocks for an equal amount of wood, *i.e.,* for 40 hundred-blocks. The 48 hundred-blocks are then parcelled out evenly among 6 sets. The 8 in the hundreds place is recorded to show that 8 hundred-blocks have been placed in each of the 6 sets, and the 4800 is written in the algorithm to show how many have been taken from the initial pile of blocks. (The initial pile of blocks can be called the dividend pile.) After subtracting to see how many remain in the initial pile, the resulting 18 should be compared with the 1 ten-block and 8 unit-blocks remaining to verify that the record (the algorithm) accurately describes what remains.

The next task is to parcel out ten-blocks among the 6 sets; however, the 1 ten-block cannot be partitioned among 6 sets. It is therefore necessary to exchange the 1 ten-block for an equal number—for 10 unit-blocks. Before exchanging, *have the child record in the algorithm with a zero that no ten-blocks are being partitioned* among the 6 sets. Finally, proceed to partition the 18 unit-blocks and complete the algorithm.

Paper 30 in Appendix A involves division with whole numbers. Can you find the error pattern?

Error Pattern E-F-1 *(from pages 73 and 111)*

When attempting to change a fraction to lower terms, Greg used his own cancellation procedure. Which of the instructional activities suggested by you as help for this child, are similar to the activities described?

E.
$$\frac{16}{64} = \frac{1}{4}$$

F.
$$\frac{14}{42} = \frac{1}{2}$$

Note: You will probably want to extend your diagnosis to see if the child can interpret a fraction with some form of physical representation. If not, instruction should focus first of all on the meaning or meanings of a fraction. The activities that follow are suggested with the assumption that the student has a basic understanding of the fraction idea.

1. *Emphasize prime factorizations.* Show that both the numerator and the denominator can be renamed as products. If the unique name for a number we call the prime factorization is used, common factors can be noted and the greatest common factor can be determined easily. When the fraction is rewritten with prime factorizations, the child's cancellation procedure *is* appropriate.

2. *Use the multiplicative identity to "go both ways."* Have the child use names for one of the form *n/n* when multiplying to rename a common fraction to higher terms. Then explore the question, How can we change the fraction back to the way it was? Show that both numerator and denominator can be divided by the same number without changing the value of the fraction.

3. *Interpret a fraction as a ratio of disjoint sets.* Construct disjoint sets to show both numerator and denominator and compare them. For the unit fraction ¼ and any fraction that is equivalent to ¼, the denominator is four times as great as the numerator. Have the child evaluate his written work keeping this relationship in mind.

Error Pattern E-F-2 *(from pages 74 and 112)*

The error pattern illustrated was explained: "3 goes to 1, and 4 goes to 2." Are the instructional activities you suggested similar to any of the activities described here?

E.
$$\frac{3}{4} = \frac{1}{2}$$

F.
$$\frac{2}{8} = \frac{1}{4}$$

Note: It may be wise to extend the diagnosis to determine if the student is able to interpret a fraction as parts of a region or a set. If the student cannot, instruction should begin with a concept of a fraction. The following activities assume the student has a basic understanding of a fraction even though a mechanical rule for changing a fraction to simplest terms was adopted.

1. *Use fractional parts of regions.* Begin with reference to the unit, then show the given number with fractional parts. Do *not* restrict the instruction to pie shapes, but use rectangular shapes as well. Pose the question, "Can we use larger parts to cover what we have exactly?" Record several "experiments" and note the ones that are already in simplest terms. Then look for a mathematical rule for changing, *i.e.,* dividing both numerator and denominator by the same number.

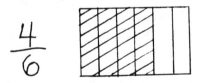

2. *Build an array with fractional parts of a set.* With discs of two colors, make a row for a given fraction.

Then, build an array by forming additional rows of discs—rows identical to the first. As each row is formed, count the columns and the discs. Record the equivalent fractions.

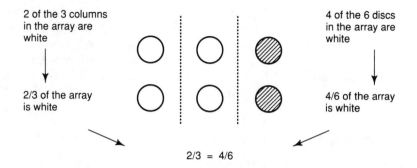

2 of the 3 columns
in the array are
white

2/3 of the array
is white

4 of the 6 discs
in the array are
white

4/6 of the array
is white

2/3 = 4/6

3. *Look for a pattern in a list.* Present a list of *correct* examples similar to the following ones. (Equivalent fractions can be thought of as two pictures of the same number, but wearing different clothes.) Have the student look for a pattern (a mathematical rule) for changing. Test out the suggested pattern on other examples. When a correct procedure is found, use it to help determine the fractions that can be changed to simpler terms and those already in simplest terms.

$$\frac{6}{8} = \frac{3}{4}$$

$$\frac{4}{6} = \frac{2}{3}$$

4. *Make sets of equivalent fractions.* Children can do this by subdividing a region, then continuing to subdivide it again and again. Record the resulting sets of equivalent fractions in order, with the lesser numerator first; then use the sets for finding simplest terms. Look for a relationship between any one fraction and the first fraction in the set.

$$\left\{ \frac{2}{3}, \frac{4}{6}, \frac{6}{9}, \frac{8}{12}, \ldots \right\}$$

5. *Have a race.* Play a board game in which players race their pieces forward along a track made up of sections, each of which is partitioned into twelfths. Players roll special dice, draw cards with fraction numerals, or draw unit regions for fractions. Possible fractions include ½, ⅔, ⅚, ¾, and so forth. Moves forward are for the equivalent number of twelfths. An example of this kind of game can be found in the Fraction Bars program.[3]

Error Pattern E-F-3 *(from pages 75 and 113)*

In order to change to lowest terms, Sue divided the greater number by the lesser to determine the new numerator, and copied the greater number as the new denominator. Which of the instructional activities you suggested are among the activities described?

G. $\dfrac{3}{6} = \dfrac{2}{6}$ H. $\dfrac{6}{4} = \dfrac{1}{6}$

Note: There is some evidence that this child is only manipulating symbols in a mechanistic way and not even interpreting fractions as part of unit regions. For example, the child's statement that ⅜ = ⅔ suggests that an understanding of ⅜ or ⅔ as parts of a unit just is not present, or, if it is, it is a behavior associated with something like fraction pies, and it is not applied in other contexts. Also, she probably thinks of *equals* as "results in" instead of "is the same as." It may be wise to interview her to determine how she conceptualizes *fractions* and *equals* before planning instruction.

1. *Match numerals with physical or diagrammatic representations.* To encourage the interpretation of a fraction in terms of real world referents, help the child learn and reinforce two behaviors:

    a. When given a physical or diagrammatic representation for a fraction, the child writes the fraction or picks out a numeral card showing "how much." In an activity of this sort, be sure the child understands the given frame of reference, *i.e.,* the unit.

    b. When given a fractional numeral, the child makes a representation for the fraction with blocks, parts of a unit region, sets, etc., or draws an appropriate diagrammatic representation.

2. *Use similar shapes.* Introduce similar figures, and show that ratios of sides are equivalent fractions. You may want to give the child several cut-out rectangles; a colored 3 cm × 4 cm card, and several white cards—some shaped similar to the 3 cm × 4 cm card, and some not. Have the child measure the cards and

determine which are the same shape as the colored card, then see how many statements of equivalence can be written using the cards.

Possible card measurements

Reference card: 3 cm × 4 cm—colored
Similar:

6 cm × 8 cm
9 cm × 12 cm
12 cm × 16 cm

Not similar:

3 cm × 5 cm
5 cm × 8 cm
8 cm × 9 cm
10 cm × 14 cm

$$\frac{3}{4} = \frac{6}{8} \qquad \frac{6}{8} = \frac{9}{12} \qquad \frac{3}{4} = \frac{9}{12}$$

3. *Order fraction cards.* Give the child a set of cards, each with a different fraction having the same denominator. Have the child sequence the cards, thereby focusing on the fact that ⅜ ≠ ⅝. (It may be necessary to emphasize that the equality sign means "is the same as.") Encourage the child to refer to physical or diagrammatic representations as necessary to verify any decisions.

4. *Play "Can you make a whole?"* The child needs to recognize fractions that can be changed to a mixed number. Give the child a set of cards with a fraction on each card Some of the cards should have proper or common fractions; others should have improper fractions. The child plays the game by sorting individual cards into two piles: those which will "make a whole" (those equal to or greater than 1) and those which will not "make a whole." A playing partner or teacher then picks two of the sorted cards to challenge, and the child uses physical representations to prove that the challenged fractions are sorted correctly. If two children are playing, they should take turns assuming sorting and challenging roles. More specific game rules and scoring procedures (if any) can be agreed upon by the children involved.

Additional suggestions for activities are listed for pattern E-F-2. They are apt to be appropriate if the child is able to do the activities listed in terms 1 through 3.

Error Pattern A-F-1 *(from pages 76 and 114)*

What instructional activities do you suggest to help Robbie correct the error pattern illustrated? See if your suggestions are among those illustrated.

$$\text{E. } \frac{3}{4} + \frac{1}{5} = \frac{4}{9} \qquad \text{F. } \frac{2}{3} + \frac{5}{6} = \frac{7}{9}$$

Note: You should extend your diagnosis by having the child complete a variety of tasks which assess subordinate skills for adding unlike fractions. Appendix D is a hierarchy of such tasks. If you prepare a diagnostic instrument with the hierarchy as a guide, order your examples from simple to complex. You may want to put them on cards or in a learning center. Because the error pattern is so similar to the multiplication algorithm, Robbie may be a child who tends to carry over one situation into his perception of another. If so, avoid extensive practice at a given time on any single procedure.

1. *Emphasize both "horizontal" and "vertical."* When adding *un*like fractions, it is usually best to write the example vertically so the renaming can be recorded more easily. Have the child practice deciding which of the several examples should be written vertically to facilitate computation and which can be computed horizontally.

2. *Use unit regions and parts of unit regions.* Have the child first represent each addend as fractional parts of a unit region. It will be necessary for the child to exchange some of the fractional parts so they are all of the same size (same denominator). The fractional parts can then be used to determine the total number of units. This procedure should be related step-by-step to the mechanics of notation in a written algorithm, probably as an example written vertically so the renaming can be noted more easily.

3. *Contrast ratio situations.* The algorithm used by Robbie *is* appropriate when adding win-loss ratios for different sets of games. (If he wins 6 of 8 games, then he wins 3 of 4 more games, altogether he has won 9 of 12 games.) Describe fraction (ratio) situations for the child, and have him decide which ones require a common denominator for addition.

4. *Discuss counting as a strategy.* Show how counting is appropriate when the denominators are the same, but is not appropriate when they are different.

5. *Estimate answers before computing.* This may require some practice locating fractions on a number line and ordering fractions written on cards. Use phrases

like "almost a half" and "a little less than one" when discussing problems. In example F, more than a half is added to a little less than one. The result should be about one and a half.

After examining the research, Suydam states: "It seems apparent that we need to shift emphasis from having students learn rules for operations on fractions to helping them develop a conceptual base for fractions."[4]

Error Pattern A-F-2 *(from pages 77 and 114)*

How would you help a student such as Dave who uses the error pattern illustrated? Are your suggestions included among the activities described?

E. $\dfrac{1}{3} + \dfrac{3}{5} = \dfrac{4}{15}$ F. $\dfrac{3}{8} + \dfrac{4}{5} = \dfrac{7}{40}$

1. *Stress that adding requires common denominators.* It makes sense to add the numerators only if the denominators are *already* the same. Explain that the reason we sometimes multiply denominators is to find a number we can use as the denominator for both fractions. Shift the focus away from getting an answer to two questions:

- What number can I use for both denominators?
- What equivalent fractions use that denominator?

2. *Find fractions for adding numerators.* Display an arrangement of fractions with different denominators; include several fractions for each denominator. Have the child select two fractions with the same denominator, add the numerators mentally, then state the sum. Thereby, you emphasize the need for fractions to have like denominators before you add them in this way.

3. *Use a number line to focus on equivalent fractions.* Prepare a number line from zero to one, with rows of labels for halves, thirds, fourths, etc. Have the child find and state the many fractions for the same point. State that each point shows a number, and the fractions are different names for the same number; when we find an equivalent fraction we are finding a different name for the same number.

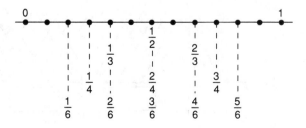

4. *Estimate answers before computing.* Consider each fraction. Is it closest to zero, to one-half, or to one? (The child may need to check a number line for reference at first.) In example E, one number is a bit less than one-half; the other number is a bit more than one-half. Their sum should be about one, rather than 4/15—which is closer to zero.

Error Pattern A-F-3 *(from pages 78 and 115)*

You have described at least two instructional activities you think would help Allen or any child with the difficulty illustrated. Are any of your suggestions among those listed?

D. $\dfrac{1}{4} + \dfrac{1}{5} = \dfrac{5}{9} + \dfrac{4}{9} = \dfrac{9}{9}$

E. $\dfrac{2}{5} + \dfrac{1}{2} = \dfrac{2}{7} + \dfrac{10}{7} = \dfrac{12}{7}$

Note: Doctoring up an erroneous mechanical procedure by trying to substitute other purely mechanical procedures frequently results in a further confusion of arbitrary and meaningless procedures. Instruction should help the student use procedures that make sense to him.

1. *Find the l.c.m. (lowest common multiple) of two whole numbers.* To help the child find the least common denominator for two fractions, work separately with the denominators as if they were whole numbers. First, make sure the child can generate sets of multiples for each whole number. If the student is able to identify the intersection set of two given sets, he will be able to find a set of *common* multiples. Finally, he can note which common multiple is *least* in value. This least common multiple is the most useful common denominator for adding the two fractions. To reinforce this skill, give the student examples where he only finds the least common denominator.

$$\dfrac{2}{3} + \dfrac{1}{4} = \boxed{} + \boxed{}$$

2. *Use the property of one for multiplication and the idea of many names for a number.* When the student can determine the least common denominator he will probably need specific help in changing one fraction to an equivalent fraction with a specified denominator. This is a specific skill that can be developed apart from the larger example.

$$\frac{2}{3} = \frac{\Box}{12}$$

If the child knows or can be taught how to multiply simple fractions, the property of one can be applied in renaming. The question posed is *"Which* name for one is useful?"

The useful name
for one

$$\frac{2}{3} \times 1 = \frac{\Box}{12} \qquad \frac{2}{3} \times \frac{\triangle}{\triangle} = \frac{\Box}{12} \qquad \frac{2}{3} \times \frac{4}{4} = \frac{\Box}{12}$$

Clue: $3 \times \triangle = 12$

3. *Change to a vertical algorithm.* A vertical algorithm permits the student to write simple equivalence statements for each renaming of a fraction.

$$\frac{2}{3} = \frac{2}{3} \times \frac{4}{4} = \frac{8}{12}$$
$$+\quad \frac{1}{4} = \frac{1}{4} \times \frac{3}{3} = \frac{3}{12}$$
$$\frac{11}{12}$$

Error Pattern A-F-4 *(from pages 79 and 116)*

Robin uses the error pattern illustrated. Are your suggestions for helping Robin among the activities described?

$$\text{D.} \quad \frac{3}{4} = \frac{3}{4} \qquad\qquad \text{E.} \quad \frac{4}{5} = \frac{4}{20}$$

$$+ \frac{1}{2} = \frac{1}{4} \qquad\qquad + \frac{1}{4} = \frac{1}{20}$$

$$\frac{}{\;\;\frac{4}{4}\;\;} \qquad\qquad \frac{}{\;\;\frac{5}{20}\;\;}$$

Note: Apparently Robin can find the least common denominator. She can also add like fractions. Corrective instruction should focus on the specific process of changing a fraction to higher terms, to a fraction with a designated denominator; e.g., $\frac{3}{4} = \frac{?}{20}$.

1. *Show that two fractions are or are not equal.* The equals sign tells us that numerals or numerical expressions on either side are names for the same number (the same fractional part, the same point on a number line). Help Robin find ways to tell if two different fractions name the same number; have her "prove" in *more than one way* that both fractions show the same amount. Examples of varied procedures that can be used include: stacking fractional parts of a unit region, using a number line which is labelled with different fractions (halves, thirds, fourths, etc.), finding a name for one (n/n) that could be used to change one of the fractions to the other, and for $a/b = c/d$ showing that $ad = bc$.

2. *Use the multiplicative identity.* Emphasize the role of one by outlining n/n with the numeral "1" as illustrated. Have the child note that *both* terms of a fraction are multiplied, and therefore, both terms in the new fraction are different than the original fraction.

$$\frac{3}{4} = \frac{3 \times \boxed{2}}{4 \times \boxed{2}} = \frac{6}{8}$$

3. *Use games involving equivalent fractions.* Have the child play games in which equivalent fractions are matched, possibly adaptations of rummy or dominoes. Such games provide an excellent context for discussing how to determine if two fractions name the same number.

4. *Use a shield.* Within the algorithm, use a shield as illustrated to help the child focus on the task of changing a fraction to higher terms. Help the child see that the procedure for changing a fraction is the same as that used *within* this algorithm.

> Appendix A includes other papers that involve adding with fractions. See if you can find the error patterns.

Error Pattern S-F-1 *(from pages 80 and 117)*

Are your suggestions for helping Andrew with the difficulty illustrated, among the suggestions listed?

E.
$$5\frac{1}{5}$$
$$-3\frac{3}{5}$$
$$\overline{2\frac{2}{5}}$$

F.
$$1$$
$$-\frac{1}{3}$$
$$\overline{1\frac{1}{3}}$$

Note: You will want to interview the child and have him think out loud as he works similar examples. Does the child question the reasonableness of his answers? In example F, the result is greater than the sum (minuend).

1. *Use fractional parts of a unit region.* Interpret the example as "take-away" subtraction and use fractional parts to show *only* the sum. If the child subtracts the whole numbers first, demonstrate that this procedure does not work; not enough remains so the fraction can be subtracted. Conclude that the fraction must be subtracted first. Have the child exchange one of the units for an equivalent set of fractional parts in order to take away the quantity indicated by the subtrahend.

2. *Use crutches to facilitate renaming.* Record the exchange of fractional parts (suggested above) as a renaming of the sum. The sum is renamed so the fraction can be subtracted easily.

$$4\frac{6}{5}$$
$$\cancel{5}\,\cancel{1}$$
$$-3\frac{3}{5}$$
$$\overline{\rule{0pt}{1em}\quad}$$
$$1\frac{3}{5}$$

3. *Practice specific prerequisite skills.* Without computing, the child can decide which examples require renaming and which do not. The skill of renaming a mixed number in order to subtract can also be practiced. You will get ideas for other prerequisite skills from Appendix D, because most of the tasks in the addition hierarchy found there apply equally to subtraction.

Error Pattern S-F-2 *(from pages 81 and 117)*

Do you find, among the suggestions listed, your suggestions for helping Chuck who has the difficulty illustrated?

E. $\quad 6\frac{2}{3} - 3\frac{1}{6} = 3\frac{1}{3}$ $\qquad$ F. $\quad 4\frac{5}{8} - 1\frac{3}{4} = 3\frac{2}{4}$

Note: Extended diagnosis is probably wise. Most of the subordinate skills suggested in the addition hierarchy found in Appendix D apply equally to subtraction.

1. *Distinguish between "horizontal" and "vertical."* When subtracting with unlike fractions and with mixed numerals, it is usually best to write the example vertically so the renaming can be recorded more easily. Have the child practice deciding which of several examples should be written vertically to facilitate computation and which can be computed horizontally.

2. *Use fractional parts of a unit region.* Have the child, or a group of children, use fractional parts to show *only* the sum, then set apart the amount indicated by the known addend. Students will soon discover that it is necessary to deal with the fraction before the whole number. They often need to exchange in order to set apart the amount required. This task can help children relate problems more adequately to the operation of subtraction. However, it can become a cumbersome procedure, so choose examples carefully. (An appropriate example might be $3\frac{1}{6} - 1\frac{2}{3}$.) Step-by-step, relate the activity with fractional parts to the vertical algorithm.

3. *Reteach and/or practice specific prerequisite skills.* Consider the tasks listed in Appendix D. Suggestions for instructional activities have already been described for many of these. For example, for Error Pattern A-F-3 activities are suggested that are appropriate for subtraction as well as for addition of unlike fractions. These activities are concerned with least common multiples, the identity element, and the like.

Error Pattern S-F-3 *(from pages 82 and 118)*

Are your suggestions for helping Ann with the difficulty illustrated among the suggestions listed?

E. $6\frac{2}{3} - 3\frac{1}{6} = 3\frac{1}{3}$ F. $4\frac{5}{8} - 1\frac{3}{4} = 3\frac{2}{4}$

Note: Corrective instruction should focus on changing a mixed number to an equivalent mixed number, one in which the fraction has a specified denominator. Ann will need to understand *why* it is frequently necessary to rename mixed numbers with the algorithm.

1. *Prove that mixed numbers are or are not equal.* Emphasize that the whole number and fraction together constitute *a* mixed number, and that "equals" written between two mixed numbers says that they name the same number

(show the same amount, name the same point on a number line). Two mixed numbers can be shown to be equal with unit regions and fractional parts, or with an appropriately labelled number line.

2. *Make many names for a mixed number.* With the help of an aid such as unit regions and fractional parts, have the child generate as many names as possible for a given mixed number. For example:

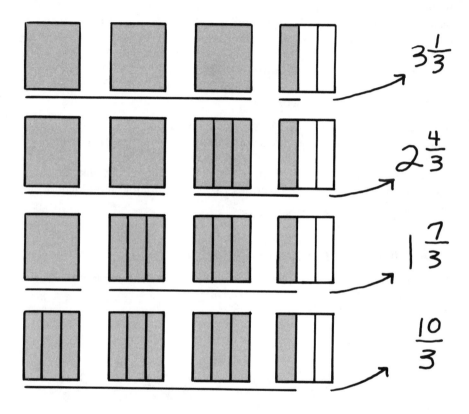

Then, for an example like $3\frac{1}{3} - 1\frac{2}{3} = \square$ ask, "Which is the most useful name for $3\frac{1}{3}$?"

3. *Point to different trading patterns.* Liken the regrouping within this algorithm to the regrouping done with whole numbers when different bases are used. Chip trading activities with varied trading rules have a structure similar to the process of changing a mixed number to an equivalent mixed number.

Appendix A includes other papers that involve subtracting with fractions. See if you can find the error patterns.

Error Pattern M-F-1 *(from pages 83 and 119)*

How would you help a student such as Dan correct the error pattern illustrated? Are the activities you described similar to any of those presented here?

E. $\frac{3}{4} \times \frac{2}{3} = 89$

F. $\frac{4}{9} \times \frac{2}{5} = 200$

Note: The child's product is most unreasonable, and continued diagnosis is wise. Does the child understand the equals sign as meaning "the same"? What kind of meaning does he associate with common fractions? Does he believe that products are *always* greater numbers?

1. *Use fractional parts of unit regions.* Interpreting an example like $3/4 \times 2/3 = ?$ as $3/4$ of $2/3 = ?$, picture a rectangular region partitioned into thirds and shade two of them. This represents $2/3$ of one. Next, partition the unit so the child can see $3/4$ of the $2/3$. What part of the unit is shown as $3/4$ of $2/3$? Be sure the child relates the answer to the unit rather than just the $2/3$. Record the fact that $3/4$ of $2/3 = 6/12$ and solve other problems with drawings. Then redevelop the rule for multiplying fractions by observing a pattern among several examples completed with fractional parts of unit regions.

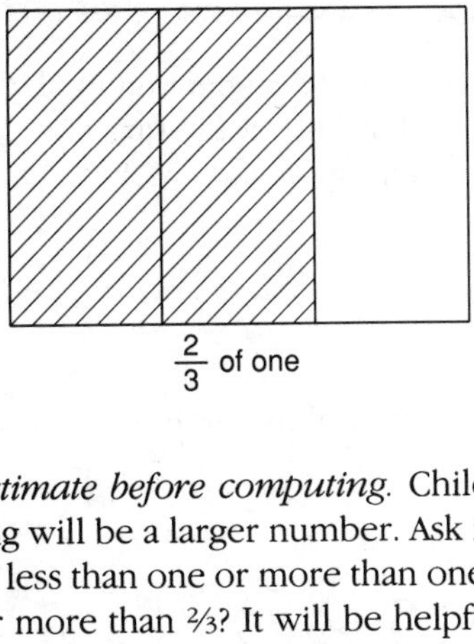

$\frac{2}{3}$ of one

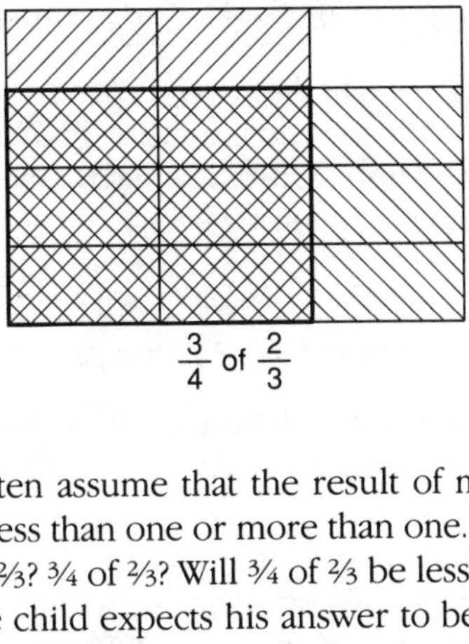

$\frac{3}{4}$ of $\frac{2}{3}$

2. *Estimate before computing.* Children often assume that the result of multiplying will be a larger number. Ask if $2/3$ is less than one or more than one. Is $1/4$ of $2/3$ less than one or more than one? $2/4$ of $2/3$? $3/4$ of $2/3$? Will $3/4$ of $2/3$ be less than $2/3$ or more than $2/3$? It will be helpful if the child expects his answer to be less than $2/3$. Of course, a child must understand fraction concepts before he can be expected to learn to estimate.

Error Pattern M-F-2 *(from pages 84 and 120)*

While she is computing, Grace may be thinking about what she has learned from someone: "Don't reason why, invert and multiply." Of course, that is exactly what she is doing.

What instructional activities do you suggest to help Grace correct the erroneous procedure illustrated? See if your suggestions are among those described.

D. $\dfrac{2}{3} \times \dfrac{3}{4} = \dfrac{2}{3} \times \dfrac{4}{3} = \dfrac{8}{9}$

E. $\dfrac{5}{7} \times \dfrac{3}{8} = \dfrac{5}{7} \times \dfrac{8}{3} = \dfrac{40}{21}$

1. *Emphasize the meaning of "equals."* *Equals* means "the same," and the expressions on either side of an equals sign should show the same number. Of course, when working with fractions it is necessary to keep in mind the fact that a given number can be expressed with many equivalent fractions, but all equivalent fractions name the same number as labels on a number line will show. Have the child examine work that has been completed and compare the expressions on both sides of the equals sign to see if they are the same number. For example, in D the ⅔ is multiplied by less than one on one side of the equals sign and by more than one on the other side.

2. *Replace computation with* yes *and* no. Note that the student multiplied correctly after the second factor had been inverted. Focus on the question, "Do I invert or not?" Give the child a mixture of multiplication and division examples to write *yes* or *no* by each.

Error Pattern M-F-3 *(from pages 85 and 120)*

How would you help Lynn correct the error pattern illustrated? Are your ideas similar to those listed?

E. $\dfrac{3}{8} \times 4 = \dfrac{12}{32}$ F. $\dfrac{5}{6} \times 2 = \dfrac{10}{12}$

1. *Make the whole number a fraction.* Show Lynn that when multiplying both terms by the same number, she is multiplying by one (in the form n/n) and not by the number given. Have her put a 1 under the whole number. Both numbers

will be fractions and the child can then use the procedure for multiplying fractions, e.g., ⅜ × ⁴⁄₁ = ¹²⁄₈.

2. *Use a number line.* On a number line which is labelled appropriately, have the child draw arrows to show the multiplication. Emphasize that the product tells how many sixths (or whatever denominator is being used). For 2 × ⅚ or for ⅚ × 2:

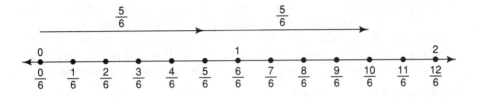

3. *Use addition.* Reverse the factors and have the child solve the addition problem suggested.

$$\frac{3}{8} \times 4 = 4 \times \frac{3}{8}$$

$$4 \times \frac{3}{8} = \frac{3}{8} + \frac{3}{8} + \frac{3}{8} + \frac{3}{8} = \frac{12}{8}$$

> Appendix A includes other papers that involve multiplying with fractions. See if you can find the error patterns.

Error Pattern D-F-1 *(from pages 86 and 121)*

What instructional activities do you suggest to help Linda correct the error pattern illustrated? See if your suggestions are among those described.

E. $\frac{4}{12} \div \frac{4}{4} = \frac{1}{3}$ F. $\frac{13}{20} \div \frac{5}{6} = \frac{2}{3}$

Note: Selection of appropriate activities will depend somewhat on which algorithm for division with fractions the child was taught originally.

1. *Introduce an alternative algorithm.* Complex fraction, common denominator, and invert and multiply are frequently taught algorithms for division with fractions. Introduce a procedure different from the one previously studied by the child.

2. *Discover a pattern.* Introduce the invert and multiply rule by presenting a varied selection of examples complete with correct answers, e.g., $\frac{7}{12} \div \frac{3}{5} = \frac{35}{36}$. Have the child compare the problems and answers and look for a pattern among the examples. Be sure each hypothesized rule is tested by checking it against all examples in the selection. After the pattern has been found, have the child verbalize the rule and make up a few examples to solve.

3. *Estimate answers with paper strips and a number line.* Using a number line and the measurement model for division, make a strip of paper about as long as the dividend and another about as long as the divisor. Ask how many strips the length of the divisor strip can be made from the dividend strip. For $\frac{5}{8} \div \frac{2}{3} = ?$, the answer might be "about one and a half." For example F, the estimate might be "a little less than one."

Error Pattern D-F-2 *(from pages 87 and 122)*

How would you help a student such as Joyce who uses the error pattern illustrated? Are your suggestions included among those listed?

D. $$\frac{5}{8} \div \frac{2}{3} = \frac{8}{5} \times \frac{2}{3} = \frac{16}{15}$$

E. $$\frac{1}{2} \div \frac{1}{4} = \frac{2}{1} \times \frac{1}{4} = \frac{2}{4}$$

Note: Determine whether the student consistently inverts the dividend, or alternates between the divisor and the dividend. You may also want to make sure the student has no difficulty distinguishing between right and left.

1. *Compare results.* Have the student compare inverting the dividend with inverting the divisor. Do both procedures produce the same result? Then explain that it is the divisor "there on the right" that is to be inverted. Have the student

suggest a way of remembering to invert the fraction on the right and not the other fraction when dividing. (Be careful. Remember the child in Chapter 1 who used the piano?)

2. *Use parts of a unit region.* Because inverting the dividend and inverting the divisor produce different answers, the student can use a manipulative aid to determine which result is correct. Parts of a unit region may be appropriate if the example is interpreted as measurement division. For ½ ÷ ¼ = ☐, have the child first place ½ of a unit on top of a unit region. This shows the dividend or product. Then explain that just as 6 ÷ 2 = ☐ asks "How many 2's are in 6?" so ½ ÷ ¼ = ☐ asks "How many ¼'s are in ½?" Have the student cover the ½ of a unit with ¼'s of a unit. In all, exactly *two* ¼'s are equal to ½. The correct result is two and not ¾; it is the result obtained by inverting the divisor on the right.

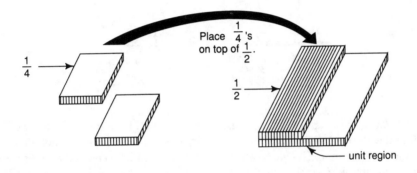

3. *Use paper strips and a number line.* These can be used as described in the previous error pattern, but used in this case to determine which fraction should be inverted for the correct result.

Papers 47–49 in Appendix A involve dividing with fractions. See if you can find the error patterns.

Error Pattern A-D-1 *(from pages 88 and 123)*

You have described two activities for helping Harold with the difficulty illustrated. Are either of your suggestions among those listed?

E.　.3
+.5
.8

F.　.7
+.7
.14

Note: Some teachers will be tempted to simply tell the child that in problems like example F the decimal point should go *between* the two digits in the sum. However, such directions only compound the problem. The child needs a greater understanding of decimal numeration and the ability to apply such knowledge. Tell students to "line up place values" when they compute with decimals; do not tell them to "line up decimal points"—that is just a result of lining up place values. Research has shown that much of the difficulty children have with decimals stems from a lack of conceptual understanding.[5]

Further diagnosis is probably wise. When the addends also include units, does the child regroup tenths as units, or does he think of two separate problems—one to the right and one to the left?

$$
\begin{array}{r}
\overset{1}{6}.7 \\
+8.5 \\
\hline
15.2
\end{array}
\qquad \text{or} \atop ? \qquad
\begin{array}{r}
6.7 \\
+8.5 \\
\hline
14.12
\end{array}
$$

1. *Use blocks or rods.* Define one size as a unit. Then have the child show each addend with blocks or rods one-tenth as large as the unit. After he combines the two sets, have him exchange tenths for a unit, if possible, so he will have "as few blocks as possible for this much wood" (or a similar expression for the particular materials used). He should compare the results of this activity with his erroneous procedure.

2. *Use a number line.* Mark units and tenths clearly on a number line and show addition with arrows. Compare the sum indicated on the number line with the sum resulting from computation.

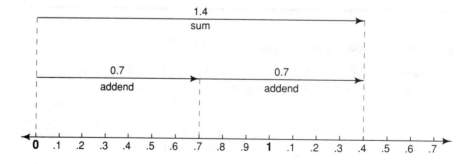

3. *Use vertically lined or cross-sectioned paper.* Theme paper can be turned 90° to use as vertically lined paper. Have the child compute using the rule that only

one digit can be placed in a column. If cross-sectioned paper is used, only one digit should be written within each square.

4. *Use metersticks.* Use metersticks to reintroduce decimals and the names for the value of each place.[6]

Papers 50 and 51 in Appendix A also involve adding with decimals. See if you can find the error patterns.

Error Pattern S-D-1 *(from pages 89 and 123)*

You described ways to help Les or other students who subtract as illustrated when they encounter ragged decimals. Are your suggestions among those listed?

D.
$$60 - 1.35 = ?$$

E.
$$24.8 - 2.26 = ?$$

$$\begin{array}{r} 60 \\ -\ 1.35 \\ \hline 59.35 \end{array}$$

$$\begin{array}{r} 24.8 \\ -\ 2.26 \\ \hline 22.66 \end{array}$$

Note: Usually the need to add or subtract decimals arises from measurement situations, and measurements should always be expressed in the same units and with the same precision if they are to be added or subtracted. Ragged decimals are inappropriate. At the same time, ragged decimals sometimes occur in situations with money (e.g., $4 − $.35). They also appear on some standardized tests, and many teachers believe students need to be taught a procedure for computing with them even if examples are somewhat contrived.

1. *Use a place value chart.* A place-value chart can be relabelled for use with decimals (e.g, tens, ones, tenths, and hundredths). For a given example, have the student first show the sum (minuend) and then work through the regrouping necessary to subtract as would be done with whole numbers. Point out that the regrouping is being done *as if* additional zeros were written to the right of the decimal point. Suggest that by affixing zeros appropriately the examples can be computed without the confusion of ragged decimals. If it will simplify things, encourage the student to affix zeros when adding as well as when subtracting. (Be sure students *affix* zeros; they do not add them.)

2. *Use base ten blocks.* For use with decimals, the unit must be defined differently from the way it is used with whole numbers, so you may choose to use blocks that are not lined. Use them as the place value chart is used.

3. *Use money.* Use pennies, dimes, dollar bills, and ten-dollar bills much as you would use a place-value chart. Stress the fact that the dollar bill is the unit; the dimes and pennies are tenths and hundredths of one dollar.

Error Pattern M-D-1 *(from pages 90 and 124)*

Illustrations of Marsha's error pattern in multiplication of decimals are shown here. Are the instructional activities you suggested to help this child among those described?

E.
$$
\begin{array}{r}
4\,0.5 \\
\times \quad .6 \\
\hline
2\,4.30
\end{array}
$$

F.
$$
\begin{array}{r}
6.7 \\
\times \quad 3 \\
\hline
2.01
\end{array}
$$

1. *Estimate before computing.* Use concepts like less than and more than in estimating the product before computing. For example E, a bit more than 40 is being multiplied by about a half. The product should be a bit more than 20. There is only one place where the decimal point could go if the answer is to be a bit more than 20. Similarly, in example F, 6.7 is between 6 and 7; therefore, the answer should be between 18 and 21. Again, there is only one place the decimal point can be placed for the answer to be reasonable. For 3.452×4.845, it can be easily seen that the product must be between 12 (i.e., 3×4) and 20 (i.e., 4×5), and there will be only one sensible place to write the decimal point.

2. *Look for a pattern.* Introduce the rule for placing the decimal point in the product by presenting a varied selection of examples complete with correct answers. Have the child compare the problems and answers and look for a pattern among the examples. Be sure the child checks her rule against all examples in the selection. When the correct rule has been established, have the child verbalize the rule and use it with a few examples she makes up herself.

Error Pattern D-D-1 *(from pages 91 and 125)*

Are your suggestions for helping Ted among the suggestions listed? He has the illustrated difficulty.

D.

$$3\overline{)2.57} \quad .852$$

$$\underline{2\,4}$$

$$1\,7$$

$$\underline{1\,5}$$

$$2$$

E.

$$.7\overline{)9.35} \quad 13.34$$

$$\underline{7}$$

$$2\,3$$

$$\underline{2\,1}$$

$$2\,5$$

$$\underline{2\,1}$$

$$4$$

1. *Label columns on lined paper.* Turn theme paper 90° and write each column of digits between two vertical lines. Then label each column with the appropriate place value. This may help discourage moving digits around mechanically. In example D, 2 hundredths is not the same as 2 thousandths.

2. *Study alternatives for handling remainders.* By using simple examples and story problems, first show that for division of *whole* numbers there are at least three different ways to handle remainders:

  a. As the amount remaining after distributing. Either a measurement or partitioning model for division can be used. The amount left over is expressed with a whole number.

$$6\overline{)3\,8\,7} \quad 64$$

$$\underline{3\,6}$$

$$2\,7$$

$$\underline{2\,4}$$

$$3$$

Answer: 64 (groups, or in each group) with 3 left over

  b. As a common fraction within the quotient expressed as a mixed number. A partitioning model for division is usually used here.

$$\begin{array}{r} 9\ 3\tfrac{1}{4} \\ 4\overline{\smash{)}3\ 7\ 3} \\ \underline{3\ 6} \\ 1\ 3 \\ \underline{1\ 2} \\ 1 \end{array}$$

Answer: 93¼ for each of the 4

c. As an indicator that the quotient should be rounded up by one, often in relation to the cost of an item. For example, pencils priced at 3 for 29¢ would sell for 10¢ each.

$$\begin{array}{r} 9 \\ 3\overline{\smash{)}2\ 9} \\ \underline{2\ 7} \\ 2 \end{array}$$

Answer: 10¢ each

Next, consider remainders for division of decimals similarly. If the remainder in example D is viewed as the amount left over after distributing, 0.02 would remain. If it is viewed as a common fraction within the quotient, the quotient would be 0.85⅔ or 0.857.

Error Pattern P-P-1 (from pages 92 and 126)

Sara solved percent problems as shown in Examples D, E, and F.

D. Brad earned $400 during the summer and saved $240 from his earnings. What percent of his earnings did he save?

$$\frac{2\ 4\ 0}{4\ 0\ 0} = \frac{x}{1\ 0\ 0}$$

Answer: 60 %

E. Barbara received a gift of money on her birthday. She spent 80% of the money on a watch. The watch cost her $20. How much money did she receive as a birthday gift?

$$\frac{2\,0}{8\,0} = \frac{x}{1\,0\,0}$$

Answer: ___$25___

F. The taffy sale brought in a total of $750, but 78% of this was used for expenses. How much money was used for expenses?

$$\frac{7\,8}{7\,5\,0} = \frac{x}{1\,0\,0}$$

Answer: ___$10.40___

Note: When you assign percent problems, ask to see all of the work done on each problem. You need to see what ratios are derived from the problem, and if the proportion itself is correctly derived. This particular student is correctly processing the proportion once it is derived and does *not* need instruction concerning cross multiplication. For this student, corrective instruction should focus on the concepts of percent, relating data in a problem to ratios (to fractions), and possibly equal ratios.

What instructional activities did you suggest to help Sara correctly solve percent problems? See if your suggestions are among the following.

1. *Use 10 × 10 squares of graph paper.* Redevelop the meaning of percent as "per hundred." Therefore, *n*% is always *n*/100. For example:

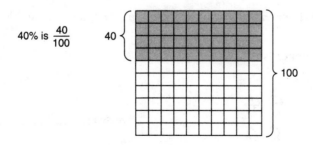

2. *Use base ten blocks.* A flat block of 100 can be partially covered with long blocks of 10 and with unit blocks to redevelop the meaning of percent.

3. *Identify what is being "counted."* Ask: "Do I know how many there are in the whole set? Do I know how many are in part of the set?" Show these numbers with a fraction:

$$\frac{\text{number in part of the set}}{\text{number in the whole set}}$$

In Problem B, for example, you are counting students. You know there are 12 students in part of the class, but you do not know how many students are in the whole class. Therefore, the fraction is

$$\frac{12}{n}$$

4. *Use number lines to show equivalent ratios.* Sketch a number line for the fraction in which both numbers are known.

$$40\% = \frac{40}{100}$$

Then sketch another line just below it for the other fraction. Because the two fractions are equal, you can align the terms of one fraction with the counterparts in the other fraction. Have the child estimate the unknown number.

$$\frac{12}{n}$$

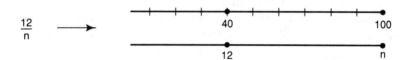

Error Pattern P-P-2 *(from pages 93 and 127)*

Steve solved percent problems as shown in Examples D, E, and F.

D. What number is 80% of 54?

```
      5 4
  x  .8 0
  ────────
      0 0
    4 3 2
  ────────
    4 3.2 0
```

Answer: __43. 2__

E. Seventy is 14% of what number?

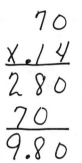

Answer: ___9.8___

F. What percent of 125 is 25?

```
   /25
 X .25
  625
 250
31.25
```

Answer: __31.25__

Note: Emphasizing a rule like "percent times a number equals percentage" is not likely to be helpful because this is actually the rule Steve is attempting to apply. Many children find it difficult to identify the three types of percent problems; they also confuse the terms *percent* and *percentage*. Instead, it may be helpful to develop a strategy that is basically the same for all three types of percent problems—possibly the proportion method.

What activities did you suggest to help Steve correctly solve percent problems of different types? See if your suggestions are among those that follow.

1. *Show equal fractions.* Write a proportion for each problem: one fraction equal to another. With one fraction show what the problem tells about percent, and with the other show what the problem tells about the number of things. Use *n* whenever you are not told a number.

For example: 70 is 14% of what number?

$$\text{percent} \quad \left\{ \frac{14}{100} = \frac{70}{n} \right. \begin{array}{l} \leftarrow \text{part of the amount} \\ \leftarrow \text{the whole amount} \end{array}$$

2. *Cross multiply.* After the student is able to write a proportion for a percent problem, suggest that the two products indicated by an *X* are equal. Have the child supply several pairs of fractions known to be equal, and cross multiply each pair to see if the rule holds for them. Then, apply cross multiplication to percent problems.

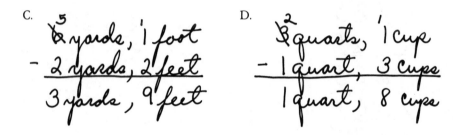

Error Pattern S-M-1 *(from pages 95 and 128)*

You have suggested instructional activities for helping Margaret, who is using the error pattern illustrated. Are any of your suggestions among those listed?

C.
$$\begin{array}{r} \overset{5}{6} \text{ yards, 1 foot} \\ - 2 \text{ yards, 2 feet} \\ \hline 3 \text{ yards, 9 feet} \end{array}$$

D.
$$\begin{array}{r} \overset{2}{3} \text{ quarts, 1 cup} \\ - 1 \text{ quart, 3 cups} \\ \hline 1 \text{ quart, 8 cups} \end{array}$$

1. *Use measuring devices.* First, have the child show the minuend with measuring devices. In example C it could be shown with yardsticks and foot rulers. Then have the child take away as much length (volume, etc.) as is suggested by the subtrahend. In the process it will be necessary to exchange. Be sure to point out that the exchanges are not always with ones and a ten; many other kinds of exchanges occur with measurement situations.

2. *Regroup in many different number bases.* Use multibase blocks, chip trading activities, place-value charts, or sticks and bundles of sticks to learn to regroup in different number bases. A game-rule orientation in which the rule for exchanging changes from game to game may help the child generalize the regrouping pattern. For base four games and activities the rule would be "Exchange a 4 for ones"; in base twelve games the rule would be "Exchange a 12 for ones," etc. Follow such activities with computation that involves measurement. Help the child connect regrouping within computation, with exchanges that were encountered when working in other number bases.

3. *Identify number base relationships.* Give the child several examples of computation that involve measurement. Have him determine the number base relationship for each example, and also state the rule for exchanging (or

regrouping) he would use when computing. (Pairs of children could work on this activity together.)

CONCLUSION

Diagnosis is a continuous process. It continues even during instructional activity as you observe a child at work and note patterns.

As you help each child learn computation procedures, focus on concepts. Help students make a habit of asking, "Is it reasonable?" Continuously emphasize estimation. And help students monitor their own learning, possibly by keeping a journal.

Be alert to how you and children use language, and to specific meanings each child associates with the words and phrases you use during diagnosis and instruction. Be sure students make sense of manipulatives they use; make sure each child can eventually connect what they observe when using manipulatives to the written procedures they are learning.

During instruction, make sure each child is aware of his strengths. Help the child take note of progress as it is made. Proceed in very small steps, if necessary, to ensure successful experiences.

It is important that your sessions with the child be varied and include games and puzzles which are enjoyable. Build on the child's strengths. As confidence is gained and as activities associated with mathematics become enjoyable, the child will be much more open to your continued efforts to correct specific difficulties.

ENDNOTES

1. Patricia S. Davidson, Grace K. Galton, and Arlene W. Fair, *Chip Trading Activities, Book I* (Fort Collins, CO: Scott Resources, Inc., 1972), available from suppliers of materials for elementary school mathematics. This spiral bound book of instructions is part of the materials called *Chip Trading Activities,* a sequence of games, problems, and other activities involving the trading of colored chips. The various activities emphasize concepts of place value, patterns of numeration, decimal notation, regrouping in addition and subtraction, the multiplication and division processes, and numeration systems other than base ten.
2. Lauren B. Resnick is similarly concerned that children be enabled to link the semantics of the base system with the semantics of the algorithm. See her "Syntax and Semantics in Learning to Subtract" in Thomas P. Carpenter, James M. Moser, and Thomas A. Romberg (eds.), *Addition and Subtraction: A Cognitive Perspective* (Hillsdale, NJ: Lawrence Erlbaum, 1982), 136–55.
3. Albert B. Bennet, Jr., and Patricia S. Davidson, *Fraction Bars* (Ft. Collins, CO: Scott Resources, Inc., 1973). Step-by-step teacher's guides are available. The materials can be purchased from many supply houses for elementary school mathematics.
4. Marilyn N. Suydam, "Fractions," *The Arithmetic Teacher* 31, no. 7 (March 1984): 64.
5. Thomas P. Carpenter et al., "Decimals: Results and Implications from National Assessment," *The Arithmetic Teacher* 28, no. 8 (April 1981): 34–37.
6. See Robert Ashlock, "Introducing Decimal Fractions with the Meterstick," *The Arithmetic Teacher* 23, no. 3 (March 1976):201–6.

Selected References

Annotated references that follow are listed under two primary headings: diagnosis and instruction. However, many references listed under diagnosis have implications, and even suggestions, for instruction. Conversely, the discerning reader will find implications for diagnosis while reading items listed under instruction.

REFERENCES FOCUSING ON DIAGNOSIS

ASHLOCK, ROBERT R. "Testing Understanding of Concepts with Paper-and-Pencil Items," in *Focus on Learning Problems in Mathematics* 9 (Fall 1987): 49–54. The author illustrates three types of items: word statements, symbolizations, and portrayals. The limitations of each are discussed.

BACKMAN, CARL A. "Analyzing Children's Work Procedures," in Marilyn Suydam and Robert Reys, eds., *Developing Computational Skills,* 1978 Yearbook of the National Council of Teachers of Mathematics (Reston, VA: The Council, 1978), 177–95. Beckman describes and illustrates errors within the several categories of a classification system. Suggestions for instruction are included.

BAROODY, ARTHUR J. "Children's Difficulties in Subtraction: Some Causes and Cures," in *The Arithmetic Teacher* 32 (November 1984): 14–19. Baroody emphasizes the need for us to diagnose a child's informal subtraction procedures so we can plan specific remedial instruction.

BEATTIE, IAN D., and JANET K. SCHEER. "Diagnosis and Remediation of Iconic Knowing in Elementary Mathematics" in *Focus on Learning Problems in Mathematics* 7 (Spring 1985): 23–28. The authors argue that iconic knowing is a key element in meaningful instruction and assessment of that learning, and they illustrate ways stamps of iconic representations can be used with children during diagnosis and remediation.

BEATTIE, JOHN, and BOB ALGOZZINE. "Testing for Teaching," in *The Arithmetic Teacher* (September 1982): 47–51. The authors illustrate tests that highlight error patterns, and describe remedial activities.

BEHR, MERLYN, STANLEY ERLWANGER, and EUGENE NICHOLS. "How Children View the Equals Sign," in *Mathematics Teaching* 92 (September 1980): 13–15. Interviews with six- and seven-year-old children reveal varied concepts associated with the equals sign in number sentences.

BEHR, MERLYN J., et al. "Order and Equivalence of Rational Numbers: A Clinical Teaching Experiment," in *Journal for Research in Mathematics Education* (November 1984):

323–41. The authors describe both correct and erroneous strategies used by students, with implications for instruction.

BOOTH, LESLEY R. "Children's Difficulties in Beginning Algebra," Chapter 3 in *The Ideas of Algebra, K–12,* Arthur Coxford and Albert Schulte, eds., 1988 Yearbook of the National Council of Teachers of Mathematics (Reston, VA: The Council, 1988), 20–32. The author uses sample dialogue to illustrate difficulties some students have in conceptualizing the use of variables.

BRASWELL, JAMES S., and ALICIA A. DODD. *Mathematics Tests Available in the United States and Canada* (Reston, VA: National Council of Teachers of Mathematics, 1988). Tests are classified by targeted grade levels and courses. Descriptions and information on availability are given.

BRIGHT, GEORGE W. "Computers for Diagnosis and Prescription in Mathematics," in *Focus on Learning Problems in Mathematics* 9 (Spring 1987): 29–41. Bright describes two general approaches to computer diagnosis of error patterns.

BROWN, JOHN SEELY, and KURT VANLEHN. "Towards a Generative Theory of 'Bugs'," in Thomas P. Carpenter, James M. Moser, and Thomas A. Romberg, eds., *Addition and Subtraction: A Cognitive Perspective* (Hillsdale, NJ: Lawrence Erlbaum, 1982): 117–35. The authors describe their research with computer systems for diagnosing systematic student errors (specifically, BUGGY and DEBUGGY) and provide explanations for error patterns—why and how they are formed.

BRUECKNER, LEO J. *Diagnostic and Remedial Teaching in Arithmetic* (Philadelphia: John C. Winston, 1930). A classic in the area of diagnosis and treatment of disabilities in mathematics. Much of this book is based on early studies of the errors pupils make in computation.

BRUMFIELD, ROBERT D., and BOBBY D. MOORE. "Problems with the Basic Facts May Not Be the Problem," in *The Arithmetic Teacher* 33 (November 1985): 17–18. The authors classify observed errors in addition and subtraction and suggest remediation that focuses on types of errors.

BURTON, GRACE M. "Young Childrens' Choices of Manipulatives and Strategies for Solving Whole Number Division Problems," in *Focus on Learning Problems in Mathematics* 14 (Spring, 1992): 2–17. Burton studied the use of manipulatives by second-grade children while solving both measurement and partition division problems. She was especially interested in determining the extent to which children select materials that match the story content of the problem situation.

CARPENTER, THOMAS P. et al. "Decimals: Results and Implications from National Assessment," in *The Arithmetic Teacher* 28 (April 1981): 34–47. The authors found that much of the difficulty with decimals lies in a lack of conceptual understanding.

CARPENTER, THOMAS P. et al. "Subtraction: What Do Students Know?" in *The Arithmetic Teacher* 22 (December 1975): 653–57. Selected data from the 1972–73 NAEP mathematics assessment are presented, and errors are discussed.

CAWLEY, JOHN F., ed. *Cognitive Strategies and Mathematics for the Learning Disabled* (Rockville, MD: Aspen Systems, 1985). A guide for constructing a classroom test of mathematics learning by disabled children is followed by suggestions for instruction.

CAWLEY, JOHN F., ed. *Practical Mathematics: Appraisal of the Learning Disabled* (Rockville, MD: Aspen Systems, 1985). This book is an in-depth treatment of assessment and diagnosis of learning disabled students.

CHARLES, RANDALL, FRANK LESTER, and PHARES O'DAFFER. *How to Evaluate Progress in Problem Solving* (Reston, VA: National Council of Teachers of Mathematics, 1987). The authors describe evaluation techniques for various levels of instruction and suggest ways to use the results.

CLARK, H. CLIFFORD. "How to Check Elementary Mathematics Papers," in *The Arithmetic Teacher* 34 (September 1986): 37–38. Clark suggests a way to have more time for careful diagnosis and evaluation of those students that need it.

CLARKE, DAVID J. "Activating Assessment Alternatives in Mathematics," in *The Arithmetic Teacher* 39 (February 1992): 24–29. Clarke describes several assessment and record-keeping procedures for classroom situations, including annotated class lists, work folios, student-constructed tests, and student self-assessment.

CLEMENT, JOHN. "Algebra Word Problem Solutions: Thought Processes Underlying a Common Misconception," in *Journal for Research in Mathematics Education* 13 (January 1982): 16–30. Clement's research identifies two conceptual sources of reversal errors: one syntactic and the other semantic.

CLEMENTS, M. A. "Analyzing Children's Errors on Written Mathematical Tasks," in *Educational Studies in Mathematics* 11 (1980). Newman's hierarchy of error causes is emphasized. Data are included on errors made by children on verbal arithmetic problems.

CLEMENTS, M. A. "Careless Errors Made by Sixth-Grade Children on Written Mathematical Tasks," in *Journal for Research in Mathematics Education* 13 (March 1982): 136–44.

COOPER, MARTIN. "The Dependence of Multiplicative Reversal on Equation Format," in *The Journal of Mathematical Behavior* 5 (August 1986): 115–20. Cooper found that whether a multiplication sign was actually used or not made a significant difference even among secondary students.

COX, L. S. "Diagnosing and Remediating Systematic Errors in Addition and Subtraction Computations," in *The Arithmetic Teacher* 22 (February 1975): 151–57. Cox emphasizes that teachers must look for patterns in the work they collect from pupils having difficulty with computation. She describes three categories of errors which can be noted.

COX, L. S. "Systematic Errors in the Four Vertical Algorithms in Normal and Handicapped Populations," in *Journal for Research in Mathematics Education* 6 (November 1975): 202–20. The author documents the fact that many children use specific erroneous procedures. Data on the frequency of selected error patterns is included.

DAVIS, ROBERT B. "The Convergence of Cognitive Science and Mathematics Education," in *The Journal of Mathematical Behavior* 5 (December 1986): 323–29. In this stimulating article Davis discusses types of learning, knowledge, and memory storage. Implications for diagnosis abound.

DAVIS, ROBERT B. "Learning Mathematical Concepts: The Case of Lucy," in *The Journal of Mathematical Behavior* 4 (October 1985): 135–53. Davis focuses on one secondary student having difficulty in a calculus course, because he believes that teachers can learn the most from students having difficulty.

DECORTE, ERIK, and LIEVEN VERSCHAFFEL. "Beginning First Graders' Initial Representation of Arithmetic Word Problems," in *The Journal of Mathematical Behavior* 4 (April 1985): 3–21. The authors analyze children's errors in word problem solving, and point to the main difficulty as construction of a mental problem representation that reflects the correct understanding of the story.

DELOCHE, G., and X. SERON, eds. *Mathematical Disabilities* (Hillsdale, NJ: Lawrence Erlbaum, 1987). The authors have assembled articles on acalculia and number-processing disorders. The neuropsychological approach to mathematical cognition is emphasized.

ENGELHARDT, JON M. "Analysis of Children's Computational Errors: A Qualitative Approach," in *British Journal of Educational Psychology* 47 (1977): 149–54. Errors are classified into eight types: basic fact error, grouping error, inappropriate inversion, incorrect operation, defective algorithm, incomplete algorithm, identity error, and zero error. The distribution of errors among the types is examined and tentative generalizations are presented.

ENGELHARDT, JON M. "Using Computational Errors in Diagnostic Teaching," in *The Arithmetic Teacher* 29 (April 1982): 16–19. The author focuses on diagnosis and remediation of four types of errors: mechanical, careless, conceptual, and procedural. Research related to conceptual errors is described.

ENRIGHT, BRIAN E. *Basic Mathematics: Detecting and Correcting Special Needs* (Boston: Allyn & Bacon, 1989), 160 pp. This is a resource for detecting and correcting mathematical needs of exceptional students. Analysis of error patterns is stressed, and suggestions are made for corrective instruction. Flowcharts for algorithms are included. It is suggested that students use the flowcharts for self-monitoring.

FENNELL, FRANCIS. "Diagnostic Teaching, Writing and Mathematics," in *Focus on Learning*

Problems in Mathematics 13 (Summer 1991): 39–50. Fennell explores the role of student writing for diagnosing student knowledge and skills. A math pals letter project is described, and many suggestions are made for diagnostic teaching.

FENNELL, FRANCIS. *Elementary Mathematics Diagnosis and Correction Kit* (West Nyack, NY: Center for Applied Research in Education, 1981). Chapters on diagnosing achievement and attitudes are followed by chapters on correcting difficulties in major areas of elementary school mathematics.

GAROFALO, JOE, and FRANK K. LESTER, JR. "Metacognition, Cognitive Monitoring, and Mathematical Performance," in *Journal for Research in Mathematics Education* 16 (May 1985): 163–76. The authors define and discuss metacognition in relation to mathematical performance. A framework for the study of performance is included, as well as a discussion of the role of metacognition in instruction.

GINSBURG, HERBERT. *Children's Arithmetic: The Learning Process* (New York: D. Van Nostrand, 1977). The author describes the process of learning arithmetic by young children, emphasizing competencies that often go unnoticed. Most chapters close with a statement of useful principles.

GLENNON, VINCENT J., and JOHN W. WILSON. "Diagnostic-Prescriptive Teaching," in *The Slow Learner in Mathematics,* 35th yearbook of the National Council of Teachers of Mathematics (Washington, DC: The Council, 1972): 282–318. In this classic within the literature of diagnostic-prescriptive teaching of mathematics, the authors first describe three major variables to consider when planning curriculum. They then present a model for cognitive diagnosis and prescription; the model includes a content taxonomy, psychological learning products, and behavioral indicators. Diagnostic-prescriptive procedures are discussed.

GOODSTEIN, HENRY A. "Are the Errors We See the True Errors? Error Analysis in Verbal Problem Solving," in *Topics in Learning and Learning Disabilities* (October 1981):31–45. The author explores the range of possible causes for errors in verbal problem solving among learning disabled students.

GRAEBER, ANNA O., and KAY M. BAKER. "Curriculum Materials and Misconceptions Concerning Multiplication and Division," in *Focus on Learning Problems in Mathematics* 13 (Summer 1991): 25–38. The authors analyze student texts in relation to common misconceptions regarding multiplication and division. The article has many implications for both diagnosis and instruction.

GRAEBER, ANNA O., and KAY M. BAKER. "Little into Big Is the Way It Always Is," in *The Arithmetic Teacher* 39 (April 1992): 18–19. The authors conclude that many children believe you "always take the little into the big" and "all operations are commutative." They also observe that children do not connect what they know about fractions with division of whole numbers. Suggestions for instruction in grades four and above are included.

GRAEBER, ANNA O., and DINA TIROSH. "Insights Fourth and Fifth Graders Bring to Multiplication and Division with Decimals," in *Educational Studies in Mathematics* 21 (December 1990): 565–88. The authors describe conceptions held by children which may impede their work with decimals, and they discuss implications of their findings. Students in both the United States and Israel were studied.

GRAEBER, ANNA O., and LISA WALLAE. *Identification of Systematic Errors: Final Report* (Philadelphia: Research for Better Schools, 1977). (ERIC Document Reproduction Service No. ED 139 662). For completed Individually Prescribed Instruction (IPI) tests, this study reports on the use of faulty algorithms by students in grades 1–6.

HANSEN, RANDALL S., JOAN McCANN, and JEROME L. MYERS. "Rote Versus Conceptual Emphases in Teaching Elementary Probability," in *Journal for Research in Mathematics Education* 16 (November 1985): 364–74. The authors observed patterns of errors across students that reflect the degree of their understanding of concepts under study.

HART, KATHLEEN M. *Ratio: Children's Strategies and Errors, A Report of the Strategies and Errors in Secondary Mathematics Project,* Centre for Science and Mathematics Education, Chelsea College, University of London (NFER-Nelson, 1984). Hart reports on the outcome of three years of research into the misconceptions underlying certain widespread errors in ratio and proportion among secondary school children. The researchers attempted to determine reasons for the errors students were making, and also tested specific teaching modules for helping students avoid the particular misconceptions.

HELM, HUGH, and JOSEPH D. NOVAK. *Proceedings of the International Seminar on Misconceptions in Science and Mathematics* (Ithaca, NY: Cornell University, July 1983). (ERIC Document Reproduction Service No. ED 242 553). Included in this extensive compilation of technical conference presentations are many descriptions of erroneous mathematical learnings and suggestions for instruction that might avoid such misconceptions.

INSKEEP, JAMES E., JR. "Diagnosing Computational Difficulty in the Classroom," in Marilyn N. Suydam and Robert E. Reys, eds., *Developing Computational Skills,* 1978 Yearbook (Reston, VA: The Council of Teachers of Mathematics, 1978), 163–76. Inskeep describes a procedure for teachers to follow when preparing and interpreting classroom tests of skill in computation.

JANSSON, LARS C. "Judging Mathematical Statements in the Classroom," in *The Arithmetic Teacher* 18 (November 1971): 463–66. The author alerts us to ways children often interpret statements in the mathematics lesson and suggests judgments teachers need to make when evaluating such statements.

KALIN, ROBERT. "How Students Do Their Division Facts," in *The Arithmetic Teacher* 31 (November 1983): 16–20. The author describes a diagnostic test and interview and illustrates how using the question, "How do you know?" enables the interviewer to determine the level of mastery or understanding.

KAMII, CONSTANCE, and BARBARA ANN LEWIS. "Achievement Tests in Primary Mathematics: Perpetuating Lower-Order Thinking," in *The Arithmetic Teacher* 38 (May 1991): 4–9. The authors caution that we can get misleading information from achievement tests. Limitations of achievement tests are described, and the value of interviews is highlighted.

KERSLAKE, DAPHNE. *Fractions: Children's Strategies and Errors.* (Windsor, Berkshire: NFER-NELSON, 1986). The author reports research conducted in conjunction with the Strategies and Errors in Secondary Mathematics Project at Chelsea College, University of London. Interviews, teaching experiments, and classroom trials of instructional procedures all contributed to conclusions about children's difficulties with fractions. Detailed evaluations of errors are included.

KILIAN, LAWRENCE et al. "Errors That Are Common in Multiplication," in *The Arithmetic Teacher* 27 (January 1980): 22–25. The authors report data on types of errors made as children multiply: procedural errors that involve zeros and carrying and calculation errors that involve multiplication facts of tables over five.

LANKFORD, FRANCIS G., JR. "What Can a Teacher Learn about a Pupil's Thinking through Oral Interviews?," in *The Arithmetic Teacher* 21 (January 1974): 26–32. The author compares computational strategies and examines errors. The article was republished as an Arithmetic Teacher classic in the October 1992 issue (pp. 106–111). Lankford is also the author of the study quoted in Appendix C of this book.

LAURSEN, K. W. "Errors in First Year Algebra," in *The Mathematics Teacher* 71 (1978): 194–95. The author describes some of the common errors in elementary algebra.

LEAN, GLENDON A., et al. "Linguistic and Pedagogical Factors Affecting Children's Understanding of Arithmetic Word Problems: A Comparative Study," in *Educational Studies in Mathematics* 21 (April 1990): 165–91. The authors examined children's strategies and errors when solving arithmetic word problems. For the Australian and Papua New Guinea children studied, strategies and errors were similar, with semantic structure of the questions in the word problems as the main variable determining difficulty.

LIEDTKE, WERNER. "Diagnosis in Mathematics: The Advantages of an Interview," in *The Arithmetic Teacher* 36 (November 1988): 26–29. The author includes interview strategies and specific interview protocols.

LIEDTKE, WERNER. "Learning Difficulties: Helping Young Children with Mathematics—Subtraction," in *The Arithmetic Teacher* 30 (December 1982): 21–23. The author focuses on diagnosis of difficulties with subtraction facts.

LINDQUIST, MARY MONTGOMERY. "Assessing through Questioning," in *The Arithmetic Teacher* 35 (January 1988): 16–18. The author gives examples of questions that can help us know what a child understands, but recognizes it is better not to ask some questions.

LOCKHEAD, JACK, and JOSE MESTRE. "From Words to Algebra: Mending Misconceptions," Chapter 13 in *The Ideas of Algebra, K–12,* Arthur Coxford and Albert Schulte, eds., 1988 Yearbook of the National Council of Teachers of Mathematics (Reston, VA: The Council, 1988), 127–35. The authors illustrate difficulties students have translating word problems into mathematical statements, then describe an instructional approach in which students identify inconsistencies inherent in their misconceptions.

LONG, MADELEINE J., and MEIR BEN-HUR. "Informing Learning through the Clinical Interview," in *The Arithmetic Teacher* 38 (February 1991): 44–46. The authors demonstrate the value of interviews as a tool for classroom teachers to uncover learning difficulties, even among students who get correct answers on test papers. Three components of the interview process are discussed.

MARKOVITS, ZVIA, BAT SHEVA EYLON, and MAXIM BRUCKHEIMER. "Difficulties Students Have with the Fraction Concept, Chapter 5 in *The Ideas of Algebra, K–12,* Arthur Coxford and Albert Schulte, eds., 1988 Yearbook of the National Council of Teachers of Mathematics (Reston, VA: The Council, 1988): 43–60. The authors describe difficulties and misconceptions students experience with reference to the function concept. Remedies are also suggested.

MARQUIS, JUNE. "Common Mistakes in Algebra," Chapter 27 in *The Ideas of Algebra, K–12,* Arthur Coxford and Albert Schulte, eds., 1988 Yearbook of the National Council of Teachers of Mathematics (Reston, VA: The Council, 1988): 204–5. The author uses her own test with students to find common mistakes in algebra. The test is presented.

MAURER, S. B. "New Knowledge about Errors and New Views about Learners: What They Mean to Educators and More Educators Would Like to Know," in A. H. Schoenfeld, ed., *Cognitive Science and Mathematics Education* (Hillsdale, NJ: Erlbaum, 1987), 165–87. Maurer discusses recent research on error patterns by cognitive theorists with implications for teaching and learning. The focus is on subtraction of whole numbers.

MESTRE, JOSE, and WILLIAM GERACE. "The Interplay of Linguistic Factors in Mathematical Translation Tasks," in *Focus on Learning Problems in Mathematics* (Winter 1986): 59–72. From their study the authors note the need to help students understand the difference between labels and variables.

MOVSHOVITZ-HADAR, NITSA, ORIT ZASLAVSKY, and SHLOMO INBAR. "An Empirical Classification Model for Errors in High School Mathematics," in *Journal for Research in Mathematics Education* 18 (January 1987): 3–14. The authors analyzed errors and developed an error classification system with six categories of errors.

NICKERSON, RAYMOND S., DAVID N. PERKINS, and EDWARD E. SMITH. "Problem Solving, Creativity, and Metacognition," Chapter 4 in *The Teaching of Thinking* (Hillsdale, NJ: Lawrence Erlbaum, 1985). Within this chapter of their book on teaching thinking, the authors focus on problem-solving strategies and on metacognition. Both have implications for diagnosis and instruction.

NOVILLIS, CAROL F. "An Analysis of the Fraction Concept into a Hierarchy of Selected Subconcepts and the Testing of the Hierarchical Dependencies," in *Journal for Research in Mathematics Education* 7 (May 1976): 131–44. Novillis focuses on the complexity of the fraction concept, and seeks to identify defining characteristics that can be taught to elementary school students. Many of the tasks in her analysis involve part-group and part-whole models.

OKEY, JAMES R., and JOHN McGARITY. "Classroom Diagnostic Testing with Microcomputers," in *Science Education* 66 (July 1982): 571–77. The authors discuss testing with a microcomputer and present a BASIC program designed for this purpose.

ORR, ELEANOR W. *Twice As Less: Black English and the Performance of Black Students in Mathematics and Science* (New York: NY: W. W. Norton, 1987). Orr illustrates how differences between Black English vernacular and standard English cause difficulties for Black students as they solve problems involving quantitative relationships.

QUINTERO, ANA H. "Conceptual Understanding in Solving Two-Step Word Problems with a

Ratio," in *Journal for Research in Mathematics Education* 14 (March 1983): 102–12. Quintero found that the meaning of concepts and relationships was the major source of difficulty in solving the problems; the concept of ratio was the key source of difficulty.

RADATZ, HENDRIK. "Error Analysis in Mathematics Education," in *Journal for Research in Mathematics Education* 10 (May 1979): 163–72. The author describes a scheme for classifying errors in performing mathematical tasks. The errors are categorized in terms of their apparent causes (e.g., difficulty in processing iconic information, application of irrelevant rules).

RADATZ, HENDRIK. "Students' Errors in the Mathematical Learning Process: A Survey," in *For the Learning of Mathematics—An International Journal of Mathematics Education* 1 (July 1980): 16–20. Radatz presents a history of error analysis in mathematics education and an overview of related research published in both German and English language publications.

REES, JOCELYN M. *Error Descriptions for Math Test: User's Guide.* 1986 (ERIC Document Reproduction Service No. ED 274 563). Rees has prepared a computer program for generating a description of items missed on the KeyMath Diagnostic Arithmetic Test.

RESNICK, LAUREN. "Beyond Error Analysis: The Role of Understanding in Elementary School Mathematics," in Helen N. Cheek, ed., *Diagnostic and Prescriptive Mathematics: Issues, Ideas, and Insights* (Kent, OH: Research Council for Diagnostic and Prescriptive Mathematics, 1984), 2–14. Resnick argues that a relatively small number of conceptual misunderstandings lie at the heart of many different erroneous procedures, and that instruction must link conceptual understanding to procedural skill. The research reported is within the domain of subtraction of whole numbers.

REYS, ROBERT E. "Evaluating Computational Estimation," in Harold L. Schoen and Marilyn J. Zweng, eds., *Estimation and Mental Computation.* (Reston, VA: National Council of Teachers of Mathematics, 1986 Yearbook): 225–38. The author discusses sample test items, and suggests guidelines for testing estimation.

ROBERSON, E. WAYNE, and DEBRA J. GLOWINSKI. *Computer Assisted Diagnostic Prescriptive Program in Reading and Mathematics: An Exemplary Micro-Computer Program and a Developer/Demonstrator Project,* National Diffusion Network, January 1986. (ERIC Document Reproduction Service ED 272 148). The authors have developed a data based curriculum management system which produces customized prescriptions for each participating student.

ROGAN, JOHN M. "Conceptual Mapping as a Diagnostic Aid," in *School and Science and Mathematics* 88 (January 1988): 50–59. The author found that conceptual maps produced by students provided useful information about each individual's understanding of specific concepts.

ROMBERG, THOMAS A., ed., *Mathematics Assessment and Evaluation: Imperatives for Mathematics Educators* (Albany, NY: State University of New York Press, 1992), 240 pp. This book stresses the need for valid data, and implications of the NCTM Standards for test development. Mathematics assessment alternatives are explored in the hope that authentic performance can be determined.

ROSS, SHARON. "Parts, Whole, and Place Value: A Developmental View," in *The Arithmetic Teacher* 36 (February 1989): 47–51. Ross cautions that students in their early stages of understanding of place value appear to understand more than they actually do; stages for the interpretations children assign to two-digit numerals are suggested. Implications for instruction are discussed.

SADOWSKI, BARBARA R., and DELAYNE HOUSTON MCILVEEN. "Diagnosis and Remediation of Sentence-solving Error Patterns," in *The Arithmetic Teacher* 31 (January 1984): 42–45. The author's analysis of errors in sentence-solving (with some surprises) is followed by suggestions for instruction. Students profit from identifying addends, sums, factors, and products before finding the missing number.

SAMMONS, KAY B. et al. "Linking Instruction and Assessment in the Mathematics Classroom," in *The Arithmetic Teacher* 39 (February 1992): 11–16. The authors describe and illustrate both formative and summative assessment techniques. Investigations based on problem-solving and at-home connections are included.

SCHEER, JANET K. "The Etiquette of Diagnosis," in *The Arithmetic Teacher* 27 (May 1980): 18–19. Twelve guidelines for diagnosis are presented.

SCHOENFELD, ALAN H. "Making Sense of 'Out Loud' Problem-Solving Protocols," in *The Journal of Mathematical Behavior* 4 (October 1985): 171–91. Schoenfeld argues that the circumstances in which people generate "out loud" solutions often affect those solutions in a variety of ways.

SHARMA, MAHESH C. "Dyscalculia and Other Learning Problems in Mathematics: A Historical Perspective," in *Focus on Learning Problems in Mathematics* 8 (Summer/Fall 1986): 7–45. In addition to describing scholarship which has focused through the years on persons unable to perform the operations of arithmetic, Sharma also presents an extensive reference list of special interest to anyone studying the history of this aspect of education.

SIDERS, JAMES A., JANE Z. SIDERS, and REBECCA M. WILSON. "A Screening Procedure to Identify Children Having Difficulties in Arithmetic," in *Journal for Research in Mathematics Education* 16 (November 1985): 356–63. Rate-of-response measures were found to be reliable predictors.

STEFANICH, GREG P., and TERI ROKUSEK. "An Analysis of Computational Errors in the Use of Division Algorithms by Fourth-Grade Students," in *School Science and Mathematics* 92 (April 1992): 201–5. The authors classify incorrect responses of fourth graders (division algorithm for whole numbers) and study the effects of instruction for systematic errors that are identified.

SUYDAM, MARILYN N. "Addition and Subtraction—Processes and Problems," in *The Arithmetic Teacher* 33 (December 1985): 20. Suydam summarizes research findings concerning difficulties students experience when learning to add and subtract.

SUYDAM, MARILYN N. *Evaluation in the Mathematics Classroom: From What and Why to How and Where,* Revised Edition (Columbus, OH: The Ohio State University, 1986). (ERIC Document Reproduction Service No. ED 284 717). Suydam not only helps the reader develop better paper-and-pencil tests, but she suggests other approaches to evaluation as well.

TATSUODA, KIKUMI K., et al. *A Psychometric Approach to Error Analysis on Response Patterns* (Research Report 80-3 ONR). (Urbana, IL: University of Illinois, Computer-based Education Research Laboratory, 1980). (ERIC Document Reproduction Service No. ED 195 577). The authors used a test of signed-number operations to examine response patterns on achievement tests. They used multi-dimensional binary vectors for error analysis and found that some problems were correct for the wrong reasons.

TRAFTON, PAUL. "Tests—a Tool for Improving Instruction," in *The Arithmetic Teacher* 35 (December 1987): 17–18. Trafton illustrates how data for specific items can suggest ways to improve instruction.

TRAVIS, BETTY. "Computer Diagnosis of Algorithmic Error," in *Computers in Mathematics Education,* 1984 Yearbook (Reston, VA: National Council of Teachers of Mathematics, 1984), 211–16. The author describes a research study in which diagnostic and remedial problems are addressed through computer technology.

TUCKER, BENNY F. "Seeing Addition: A Diagnosis-Remediation Case Study" in *The Arithmetic Teacher* 36 (January 1989): 10–11. Tucker provides an illustration of how diagnosis can lead to specific instructional tasks that get results.

UNDERHILL, ROBERT G., A. EDWARD UPRICHARD, and JAMES W. HEDDENS. *Diagnosing Mathematical Difficulties* (New York: Merrill/Macmillan, 1980). Models for diagnosis in clinic and classroom are presented and their implications discussed. Considerable attention is also given to diagnostic tests.

UNIVERSITY OF MARYLAND. *INSIGHTS into Secondary School Students' Understanding of Mathematics,* Anna O. Graeber and Martin Johnson, principal investigators (1991). This is an instructor's manual for a course for teachers, developed with the help of a National Science Foundation grant. The manual includes student materials, overhead masters, and several articles concerning the nature of misconceptions and ways to help students overcome them. The manual is accompanied by *Resource Bibliography for INSIGHTS,* a Hypercard disc containing an annotated bibliography.

VIRGINIA COUNCIL OF TEACHERS OF MATHEMATICS. *Mathematics Assessment for the Classroom Teacher* (Charlottesville, VA: 1983). (ERIC Document Reproduction Service No. SE 046 813). The articles include many specific examples of classroom assessment practices.

WEARNE-HIEBERT, DIANA, and JAMES HIEBERT. "Junior High School Students' Understanding of Fractions," in *School Science and Mathematics* 83 (February 1983): 96–106. The analysis of errors on specific tasks suggested that student understanding of fractions suffered not so much from being incorrect: rather, understanding was incomplete.

WILDE, SANDRA. "Learning to Write About Mathematics," in *The Arithmetic Teacher* 38 (February 1991): 38–43. Wilde illustrates the diagnostic value of writing within the mathematics classroom.

YSSELDYKE, SALVIA. *Assessment in Special and Remedial Education,* 4th edition. (Boston, MA: Houghton Mifflin, 1987). Chapter 18, in particular, is devoted to diagnostic assessment in mathematics.

REFERENCES FOCUSING ON INSTRUCTION

ALLINGER, GLENN D. "Mind Sets in Elementary School Mathematics," in *The Arithmetic Teacher* 30 (November 1982): 50–53. The author illustrates how a teacher's thoughtless use of words or examples can lead to students acquiring a "functional fixedness" that makes learning a concept more difficult.

ASHCRAFT, MARK H. "Is It Farfetched That Some of Us Remember Our Arithmetic Facts?" in *Journal for Research in Mathematics Education* 16 (March 1985): 99–105. Ashcraft argues that older children and adults actually retrieve basic facts that are stored in memory, rather than reconstructing answers as Baroody claims.

ASHLOCK, ROBERT B. "Use of Informal Language When Introducing Concepts," in *Focus on Learning Problems in Mathematics* 9 (Summer 1987): 31–36. The author illustrates situations in which more formal language can be introduced in apposition to informal language while teaching concepts.

ASHLOCK, ROBERT B., and CAROLYNN A. WASHBON. "Games: Practice Activities for the Basic Facts," in Marilyn N. Suydam and Robert E. Reys, eds., *Developing Computational Skills.* (Reston, VA: National Council of Teachers of Mathematics, 1978 Yearbook), 39–50. The use of games for practice with basic facts is discussed. Games are described in relation to the guidelines presented.

ASHLOCK, ROBERT B. et al. *Guiding Each Child's Learning of Mathematics: A Diagnostic Approach to Instruction* (New York: Merrill/Macmillan, 1983). In their cognitively oriented methods text, the authors present models for guiding both diagnosis and instruction.

BAROODY, ARTHUR J. *Children's Mathematical Thinking* (New York: Teachers College Press, 1987). Appealing to cognitive theory, Baroody argues for a meaningful approach to early mathematical instruction, with an emphasis on counting activity.

BAROODY, ARTHUR J. "Mastery of Basic Number Combinations: Internalization of Relationships or Facts?" in *Journal for Research in Mathematics Education* 16 (March 1985): 83–98. Baroody emphasizes the role of learned principles and procedures as a person attempts to recall basic number facts, and contrasts his position with that of Ashcraft.

BEATTIE, IAN D. "Modeling Operations and Algorithms," in *The Arithmetic Teacher* 33 (February 1986): 23–28. Beattie presents specific procedures for correlating manipulatives, language, and algorithms—for subtraction and division.

BLEY, NANCY, and CAROL THORNTON. *Teaching Mathematics to the Learning Disabled,* 2nd ed. (Austin, TX: Pro-Ed, 1989), 521 pp. The authors present very specific instructional procedures for teaching computation. Many of the procedures are designed to help children focus on the task at hand, or make needed distinctions.

BORASI, RAFFAELLA. "Exploring Mathematics Through the Analysis of Errors," in *For the Learning of Mathematics—An International Journal of Mathematics Education* 7 (November 1987): 2–8. Borasi recommends student errors be used as starting points for mathematical explorations that involve problem-solving and problem-posing activities. Errors can not only serve as motivational devices, but also foster a more complete understanding of mathematical content.

BORASI, RAFFAELLA. "Using Errors as Springboards for the Learning of Mathematics: An Introduction" in *Focus on Learning Problems in Mathematics* 7 (Summer and Fall 1985): 1–14. The author believes that, even apart from remediation, errors can have a positive role in a student's learning of mathematics.

BRIGHT, GEORGE W., JOHN G. HARVEY, and MARGARIETE M. WHEELER, *Learning and Mathematics Games* (Reston, VA: National Council of Teachers of Mathematics, 1985), Monograph 1 of the *Journal for Research in Mathematics Education*. The authors describe research on game-related instruction at both the elementary and secondary levels.

BURTON, GRACE M. "Focus on Affective Issues in Diagnostic and Prescriptive Mathematics," in *Focus on Learning Problems in Mathematics* 10 (Spring 1988): 55–61. Burton discusses the effects of positive and negative teacher expectations, then summarizes research as to what teachers can do to facilitate positive and realistic student self-concepts.

BURTON, GRACE M. "Teaching the Most Basic Basic," in *The Arithmetic Teacher* 32 (September 1984): 20–25. Burton describes how varied aids can be used for teaching numeration.

BURTON, GRACE M., and MARCEE J. MEYERS. "Teaching Mathematics to Learning Disabled Students in the Secondary Classroom," in *The Mathematics Teacher* 80 (December 1987): 702–747. The authors focus on ways to modify mathematics instruction for learning disabled students.

CAREY, DEBORAH A. "Number Sentences: Linking Addition and Subtraction Word Problems and Symbols," in *Journal for Research in Mathematics Education* 22 (July 1991): 266–80. Carey found that the first-grade children she studied focused on the semantic structure of word problems, rather than on part-whole relationships that indicate the operation to be used to find the missing number.

CAWLEY, JOHN F., ANNE M. FITZMAURICE HAYES, and ROBERT A. SHAW. *Mathematics for the Mildly Handicapped* (Newton, MA: Allyn & Bacon, 1988). This methods text focuses on diagnostic instruction of disabled learners.

DAVIDSON, NEIL. "Cooperative Learning in Mathematics," in *Cooperative Learning* 10 (October 1989): 2–3. The author argues that cooperative learning groups can be used with mathematics students at all age levels, and he outlines classroom procedures.

DAVIS, EDWARD J. "Suggestions for Teaching the Basic Facts of Arithmetic," Marilyn N. Suydam and Robert E. Reys, eds., *Developing Computational Skills*. (Reston, VA: National Council of Teachers of Mathematics, 1978 Yearbook), 51–60. Davis lists guidelines and illustrates how they are applied.

DAVISON, DAVID M., and DANIEL L. PEARCE. "Using Writing Activities to Reinforce Mathematics Instruction," in *The Arithmetic Teacher* 35 (April 1988): 42–45. Different categories of writing activities are illustrated, including "linguistic translation," which provides diagnostic feedback for the teacher.

DESSART, DONALD, and MARILYN SUYDAM. *Classroom Ideas from Research on Secondary School Mathematics* (Reston, VA: National Council of Teachers of Mathematics, 1983). The authors review research on teaching algebra and geometry, and highlight ideas for the classroom.

DRISCOLL, MARK J. *Research Within Reach: Elementary School Mathematics*. (Reston, VA: National Council of Teachers of Mathematics, 1981). The author summarizes research as he answers questions concerning diagnosis, remediation, algorithms, and mastery learning.

DRISCOLL, MARK J. *Research Within Reach: Secondary School Mathematics* (Reston, VA: National Council of Teachers of Mathematics, 1983). The author interprets research on teaching mathematics for the classroom teacher.

ELLERBRUCH, LAWRENCE W., and JOSEPH N. PAYNE. "A Teaching Sequence from Initial Fraction Concepts through the Addition of Unlike Fractions," in Marilyn N. Suydam and Robert E. Reys, eds., *Developing Computational Skills*. (Reston, VA: National Council of Teachers of Mathematics, 1978 Yearbook), 129–47. The authors outline a sequence for developmental teaching of fraction concepts and algorithms.

ENGLEHARDT, JON M., and VIRGINIA USNICK. "When Should We Teach Regrouping in Addition and Subtraction?" in *School Science and Mathematics* 91 (January 1991): 6–9. From their exploratory studies, the authors conclude (1) that more attention may need to be given to alternate sequences of instruction that begin with more general cases and examples that involve regrouping, and (2) the traditional emphasis on basic fact mastery and numeration concept knowledge may need to be reexamined.

FISCHER, FLORENCE E. "A Part-Part-Whole Curriculum for Teaching Numbers in the Kindergarten," in *Journal for Research in Mathematics Education* 21 (May 1990): 207–15. Fischer found that kindergarten children were more successful with addition and subtraction word problems and with place-value concepts when set-subset relationships were taught.

FUSON, KAREN C. "Issues in Place-Value and Multidigit Addition and Subtraction Learning and Teaching," in *Journal for Research in Mathematics Education* 21 (July 1990): 273–80. Fuson argues for a sequence for teaching and learning about place-value and multidigit addition and subtraction in which problems with and without trades are presented at the same time.

FUSON, KAREN C. "Research on Whole Number Addition and Subtraction," in Douglas A. Grouws, ed., *Handbook of Research on Mathematics Teaching and Learning* (New York: Maxwell Macmillan International, 1992), 243–75. In her summary of research, Fuson focuses on children's conceptual structures for the operations and for addition and subtraction multidigit computation. She also stresses the variety of addition and subtraction in real-world situations.

FUSON, KAREN C., and DIANE J. BRIARS. "Using A Base-Ten Blocks Learning/Teaching Approach for First- and Second-Grade Place-Value and Multidigit Addition and Subtraction," in *Journal for Research in Mathematics Education* 21 (May 1990): 180–206. Two studies were conducted in which base ten blocks were used to carry out steps in computation. Each step in the procedure was immediately recorded with numerals. Activities focused on four-digit numbers, but included practice with five-to-eight-digit addition and subtraction.

GRAEBER, ANNA O., and KAY M. BAKER. "Little into Big Is the Way It Always Is," in *The Arithmetic Teacher* 39 (April 1992): 18–21. The authors describe the difficulties children have when they encounter quotients of less than one, noting that texts provide relatively few examples of this type. Specific ideas for overcoming this tendency are presented for teachers at different grade levels.

GROUWS, DOUGLAS A., ed. *Handbook of Research on Mathematics Teaching and Learning* (New York: Macmillan, 1992 National Council of Teachers of Mathematics), 771 pp. This extensive volume is a major resource you should consult if you want to know what research tells us about the teaching and learning of mathematics. Articles on operations include conceptual structures required by learners, with implications for both diagnosis and instruction.

HALL, WILLIAM D. "Division with Base-Ten Blocks," in *The Arithmetic Teacher* 31 (November 1983): 21–23. The author illustrates a partitive division situation with base-ten blocks.

HARRISON, MARILYN, and BRUCE HARRISON. "Developing Numeration Concepts and Skills," in *The Arithmetic Teacher* 33 (February 1986): 18–21, 60. The authors describe game-like activities for developing specific numeration related concepts.

HAZEKAMP, DONALD W. "Teaching Multiplication and Division Algorithms," in Marilyn N. Suydam and Robert E. Reys, eds., *Developing Computational Skills.* (Reston, VA: National Council of Teachers of Mathematics, 1978 Yearbook), 96–128. The author illustrates the steps in developmental instruction for these algorithms.

HIEBERT, JAMES. "Children's Mathematics Learning: The Struggle to Link Form and Understanding," in *The Elementary School Journal* 84 (May 1984): 497–513. Hiebert stresses the need to teach so that understanding and forms (such as paper-and-pencil algorithms) are linked, and he describes ways to do this.

HOPE, JACK A., and DOUGLAS T. OWENS. "An Analysis of the Difficulty of Learning Fractions," in *Focus on Learning Problems in Mathematics* 9 (Fall 1987): 25–40. Different physical and symbolic settings facilitate different meanings for fractions. Learning fractions is more difficult than is often recognized.

HUTCHINGS, BARTON. "Low-Stress Algorithms," in Doyal Nelson, and Robert E. Reys, ed., *Measurement in School Mathematics* (Reston, VA: National Council of Teachers of Mathematics, 1976 Yearbook), 218–39. The author describes procedures for whole-number computation, algorithms that children experiencing difficulty with standard procedures have found especially useful.

KAGAN, SPENCER. "The Structural Approach to Cooperative Learning," in *Educational Leadership* 47 (December 1989/January 1990): 12–15. Kagan describes several different structures teachers can use when planning cooperative groups or teams of students, and he identifies instructional functions appropriate for each structure.

KLAUER, ELIZABETH, and ANN RULE. *Teaching Strategies in Algebra: The Effectiveness of Relating and Sequencing Algebraic Concepts,* (ERIC Document Reproduction Service, 1985, ED 266 007).

LAZERICK, BETH E. "Mastering Basic Facts of Addition: An Alternate Strategy," in *The Arithmetic Teacher* 28 (March 1981): 20–24. Lazerick describes seven ordered clusters of basic facts that can facilitate memorization.

MANNING, BRENDA H. "A Self-Communication Structure for Learning Mathematics," in *School Science and Mathematics* 84 (January 1984): 43–51. Manning defends and describes teaching students concepts of self-communication; that is, talking to themselves.

MCBRIDE, JOHN W., and CHARLES E. LAMB. "Using Commercial Games to Design Teacher-Made Games for the Mathematics Classroom," in *The Arithmetic Teacher* 38 (January 1991): 14–22. The authors offer many illustrations of how teachers can design games to provide practice for specific skills.

MCKILLIP, WILLIAM D. "Computational Skill in Division: Results and Implications from National Assessment," in *The Arithmetic Teacher* 28 (March 1981): 34–37. The author focuses on the sources of error and recommends three things for improving division computation.

MEYERS, MARCEE J., and GRACE M. BURTON. "Yes You Can . . . Plan Appropriate Instruction for Learning Disabled Students," in *The Arithmetic Teacher* 36 (March 1989): 46–51. How do specific learning disabilities affect attempts to learn mathematics? The authors provide some answers, and include suggestions for instruction.

MITCHELL, CHARLES E. "The Non-Commutativity of Subtraction," in *School Science and Mathematics* 83 (February 1983): 133–39. Many children ignore order when subtracting, and the author discusses possible explanations. The need for instruction stressing the non-commutativity of subtraction is emphasized.

MOVSHOVITZ-HADAR, NITSA, SHLOMO INBAR, and ORIT ZASLAVSKY. "Students' Distortions of Theorems," in *Focus on Learning Problems in Mathematics* 8 (Winter 1986): 49–57. The authors describe and illustrate two types of errors: distortions of antecedent and distortions of the consequent. Recommendations for prevention and instruction are included.

MOYER, MARGARET B., and JOHN C. MOYER. "Ensuring That Practice Makes Perfect: Implications for Children with Learning Disabilities" in *The Arithmetic Teacher* 33 (September 1985): 40–42. The authors recommend focusing on substeps; they use subtraction with regrouping as an example.

MYERS, ANN C., and CAROL A. THORNTON. "The Learning Disabled Child—Learning the Basic Facts," in *The Arithmetic Teacher* 25 (December 1977): 46–50. The authors emphasize that learning disabled children need to learn strategies for using what they know to figure out other facts.

PARKER, JANET, and CONNIE C. WIDMER. "Computation and Estimation," in *The Arithmetic Teacher* 40 (September 1992): 48–51. The authors make many suggestions for teaching computation within problem-solving contexts. They stress the need to use calculators when appropriate to carry out the actual computations.

PERRY, LELAND M. "Mistakes in Mathematics—Terrible or Trivial?" in *The Arithmetic Teacher* 37 (January 1990): 34–37. Perry presents questionable diagrams and other "mistakes" sometimes used in texts and tests which easily lead to misunderstanding.

POST, THOMAS R. "Fractions: Results and Implications from National Assessment," in *The Arithmetic Teacher* 28 (May 1981): 26–31. The author shows that many students use rote procedures; he recommends an emphasis on estimation with fraction operations.

RATHMELL, EDWARD C. "Concepts of the Fundamental Operations: Results and Implications from National Assessment," in *The Arithmetic Teacher* 28 (November 1980): 34–37. Rathmell focuses on the unrealized potential of the number line for representing operations.

RATHMELL, EDWARD C. "Using Thinking Strategies to Teach the Basic Facts," in Marilyn N. Suydam and Robert E. Reys, eds., *Developing Computational Skills.* (Reston, VA: National Council of Teachers of Mathematics, 1978 Yearbook): 13–38. Thinking strategies are described for use when organizing instruction in the basic facts for addition and multiplication.

RESNICK, LAUREN B. "Syntax and Semantics in Learning to Subtract," in Thomas P. Carpenter, James M. Moser, and Thomas A. Romberg, eds., *Addition and Subtraction: A Cognitive Perspective* (Hillsdale, NJ: Lawrence Erlbaum, 1982), 136–55. The author has found that although children know much about base-ten numeration system, many are unable to apply this knowledge to written arithmetic procedures. She focuses on teaching methods that will help students link place-value knowledge with the sequenced steps of computation.

REYS, ROBERT E. et al. *Computational Estimation* (Palo Alto, California: Dale Seymour Publications, 1987). Separate volumes for grades six, seven, and eight provide much instruction and practice with estimation skills. Front-end estimation, rounding, and compatible numbers are included, as are computations with whole numbers, fractions, decimals, and percents.

ROSENBAUM, LINDA et al. "Step into Problem Solving with Cooperative Learning," in *The Arithmetic Teacher* (March 1989): 7–11. The authors describe how cooperative groups of students can be incorporated into mathematics instruction in the primary grades. The benefits of cooperation rather than competition are stressed.

SADOWKSI, BARBARA. "Sentence-Solving Strategies of Elementary School Children," in *School Science and Mathematics* 85 (April 1985): 317–29. The author found that students following a strategy which is 80 percent successful for solving open number sentences were not likely to abandon it for a strategy that is 100 percent successful.

SCHROEDER, THOMAS L. "Capture: A Game of Practice, a Game of Strategy," in *The Arithmetic Teacher* 31 (December 1983): 30–31. The author describes an interesting strategy game involving basic operations. It is useful for grades 3 and up.

SHAUGHNESSY, MINA P. *Errors and Expectations* (New York: Oxford University Press, 1977). This book focuses on instruction in written language, but contains thought-provoking ideas that parallel instruction in mathematics; comments about coding, syntax, and teacher responses to student errors are examples.

SHAW, ROBERT A., and PHILIP A. PELOSI. "In Search of Computational Errors," in *The Arithmetic Teacher* 30 (March 1983): 50–51. The authors illustrate the need to conduct individual interviews in addition to paper-and-pencil diagnostic procedures.

SILVIA, EVELYN M. "A Look at Division with Fractions," in *The Arithmetic Teacher* 30 (January 1983): 38–41. Silvia describes how division with fractions can be introduced with graph paper.

SOWDER, JUDITH. "Estimation and Number Sense," in Douglas A. Grouws, ed., *Handbook of Research on Mathematics Teaching and Learning* (New York: Maxwell Macmillan International, 1992), 371–89. Included in Sowder's summary of research are sections on computational estimation, mental computation, and number sense.

STEINBERG, RUTH M. "Instruction on Derived Facts Strategies in Addition and Subtraction," in *Journal for Research in Mathematics Education* 16 (November 1985): 337–55. The authors found that rate-of-response measures with first graders were effective predictors of achievement in arithmetic over a two-year period.

SUTTON, JOHN T., and TONYA D. URBATSCH. "Transition Boards: A Good Idea Made Better," in *The Arithmetic Teacher* 38 (January 1991): 4–9. The authors describe how base-ten blocks can be used with addition and subtraction transition boards when teaching multidigit computation procedures.

SUYDAM, MARILYN N. "Improving Multiplication Skills," in *The Arithmetic Teacher* 32 (March 1985): 52. Suydam summarizes research on teaching multiplication—research which supports a meaningful approach.

SUYDAM, MARILYN N., and DONALD J. DESSART. *Classroom Ideas from Research on Computational Skills* (Reston, VA: National Council of Teachers of Mathematics, 1976). The authors summarize findings from research with reference to introducing, reinforcing, maintaining, transferring, and applying computational skills with whole numbers and fractions.

SUYDAM, MARILYN N., and ROBERT E. REYS, eds. *Developing Computational Skills* (Reston, VA: National Council of Teachers of Mathematics, 1978 Yearbook). The basic facts and computation procedures for whole numbers and fractions are all considered in this very helpful reference.

SWEETLAND, ROBERT D. "Understanding Multiplication of Fractions," in *The Arithmetic Teacher* 32 (September 1984): 48–52. The author describes how cuisenaire rods can be used to make sense out of multiplying fractions.

THOMPSON, CHARLES S., and JOHN VAN DE WALLE. "Transition Boards: Moving from Materials to Symbols in Addition," in *The Arithmetic Teacher* 28 (December 1980): 4–8; and "Transition Boards: Moving from Materials to Symbols in Subtraction," in *The Arithmetic Teacher* 28 (January 1981): 4–9. The authors describe a phased transition from the use of symbols alone.

THORNTON, CAROL A. "Emphasizing Thinking Strategies in Basic Fact Instruction," in *Journal for Research in Mathematics Education* 9 (May 1978): 214–27. The effect of teaching thinking strategies for basic facts in grades two and four is explored and data are reported to support use of the specific strategies presented.

THORNTON, CAROL A., GRAHAM A. JONES, and MARGARET A. TOOHEY. "A Multisensory Approach to Thinking Strategies for Remedial Instruction in Basic Addition Facts," in *Journal for Research in Mathematics Education* 14 (May 1983): 198–203. The authors report a pilot study involving specific addition fact strategies.

THORNTON, CAROL A., and PAULA J. SMITH. "Action Research: Strategies for Learning Subtraction Facts," in *The Arithmetic Teacher* 35 (April 1988): 8–12. The authors describe how specific activity-based subtraction fact strategies were taught as part of a teaching experiment.

THORNTON, CAROL A., and BARBARA WILMOT. "Special Learners," in *The Arithmetic Teacher* 33 (February 1986): 38–41. Thornton describes specific procedures for teaching computation to learning handicapped students.

TIERNEY, CORNELIA C. "Patterns in the Multiplication Table," in *The Arithmetic Teacher* 32 (March 1985): 36–40. The author describes activities which may help students who are learning the basic multiplication facts.

TRAFTON, PAUL R., and JUDITH S. ZAWOJEWSKI. "Teaching Rational Number Division: A Special Problem," in *The Arithmetic Teacher* 31 (February 1984): 20–22. Division with both fractions and decimals is addressed.

TUCKER, BENNY F. "The Division Algorithm," in *The Arithmetic Teacher* 20 (December 1973): 639–46. Tucker describes how the division algorithm can be introduced using the partitive model and a variety of exemplars.

UPRICHARD, A. EDWARD, and CAROLYN COLLURA. "The Effect of Emphasizing Mathematical Structure in the Acquisition of Whole Number Computation Skills (Addition and Subtraction) by Seven-and-Eight-Year-Olds: A Clinical Investigation," in *School Science and Mathematics* 77 (February 1977): 97–104. Instruction on basic facts with sums 11–18 was more effective when emphasizing structures such as place value rather than just games for drill.

USNICK, VIRGINIA. "It's Not Drill AND Practice, It's Drill OR Practice," in *School Science and Mathematics* 92 (December 1991): 344–47. Usnick draws on an analogy to make a useful distinction between drill and practice. Each is appropriate at times, but the author cautions against using drill when what is actually needed is practice.

USNICK, VIRGINIA E. "Multidigit Addition: A Study of an Alternate Sequence," in *Focus on Learning Problems in Mathematics* 14 (Summer 1992): 53–62. Usnick studied the effects of two instructional sequences for the standard addition algorithm, a traditional sequence in which examples with no regrouping are introduced and practiced before examples with regrouping, and an alternative sequence in which examples with regrouping are introduced initially. The alternative sequence was as effective as the traditional sequence.

USNICK, VIRGINIA, and JON M. ENGELHARDT. "Basic Facts, Numeration Concepts, and the Learning of the Standard Multidigit Addition Algorithm," in *Focus on Learning Problems in Mathematics* 10 (Spring 1988): 1–14. From the authors' study it is not clear that actual mastery of basic facts and relevant numeration concepts is necessary before introducing the addition algorithm.

VAN DE WALLE, JOHN, and CHARLES S. THOMPSON. "Fractions with Fraction Strips," in *The Arithmetic Teacher* 32 (December 1984): 4–9. Specific guidance is given for using colored strips to develop fraction concepts and relate them to numerals.

WILLIAMS, LINDA V. *Teaching for the Two-Sided Mind: A Guide to Right Brain/Left Brain Education.* (New York: Simon and Schuster, 1983). Williams describes instructional techniques associated with right hemisphere functioning, including evocative language, metaphor, and visual strategies such as mind maps.

YACKEL, ERNA, PAUL COBB, and TERRY WOOD. "Small-Group Interactions as a Source of Learning Opportunities in Second-Grade Mathematics," in *Journal for Research in Mathematics Education* 22 (November 1991): 390–408. The authors focus on the nature of learning in cooperative groups, and they include sample dialogues that illustrate how the learning of basic facts can be approached as problem solving.

RCDPM REFERENCES AVAILABLE FROM ERIC

The following references published by the Research Council for Diagnostic and Prescriptive Mathematics are among those available through ERIC.

BEATTIE, IAN D., and others. *Research Reports from the Seventh National Conference on Diagnostic and Prescriptive Mathematics. 1981 Research Monograph* (Kent, OH: Research Council for Diagnostic and Prescriptive Mathematics, 1982). (ED 243 709).

DENMARK, TOM, ed. *Issues for Consideration by Mathematics Educators: Selected Papers* (Kent, OH: Research Council for Diagnostic and Prescriptive Mathematics, 1978). (ED 243 693).

GLENNON, VINCENT, J. *Neuropsychology and the Instructional Psychology of Mathematics.* (Kent, OH: Research Council for Diagnostic and Prescriptive Mathematics, 1981) (ED 243 697).

HYNES, MARY ELLEN, ed. *Topics Related to Diagnosis in Mathematics for Classroom Teachers.* (Kent, OH: Research Council for Diagnostic and Prescriptive Mathematics, 1978). (ED 243 694).

HYNES, MICHAEL, C., ed. *An Annotated Bibliography of Periodical Articles Relating to the Diagnostic and Prescriptive Instruction of Mathematics* (Kent, OH: Research Council for Diagnostic and Prescriptive Mathematics, 1979). (ED 243 695).

ROMBERG, THOMAS A. *Research Reports from the Fourth and Fifth National Conferences on Diagnostic and Prescriptive Mathematics. 1980 Research Monograph.* Revised Third Ed. (Kent, OH: Research Council for Diagnostic and Prescriptive Mathematics, 1981). (ED 243 708).

SADOWSKI, BARBARA, ed. *An Annotated Bibliography Relating to the Diagnostic and Prescriptive Instruction of Mathematics,* Volume 2. (Kent, OH: Research Council for Diagnostic and Prescriptive Mathematics, 1982). (ED 243 696).

SPEER, WILLIAM R. *A Bibliography Related to the Nature of Diagnosis and Remediation in Mathematics* (Kent, OH: Research Council for Diagnostic and Prescriptive Mathematics, 1981). (ED 243 701).

SPEER, WILLIAM R. *Clinical Investigations in Mathematics Education. Thematic Address from the Fourth Annual Conference on Diagnostic and Prescriptive Mathematics* (Kent, OH: Research Council for Diagnostic and Prescriptive Mathematics, 1978). (ED 243 702).

Additional
Children's Papers and
Key to Error Patterns

On the following pages are brief excerpts from the written work of children using erroneous computational procedures. Practice the skill of identifying error patterns by finding the erroneous procedure in each of these papers. Note how students tend to manipulate symbols without really thinking about numbers and operations on numbers.

Briefly describe each error pattern, then check the key on page 216 if you wish.

PAPER 1

$$\begin{array}{r} 35 \\ +28 \\ \hline 18 \end{array} \qquad \begin{array}{r} 24 \\ +17 \\ \hline 14 \end{array} \qquad \begin{array}{r} 43 \\ +26 \\ \hline 15 \end{array}$$

Description of Pattern _____

_____

PAPER 2

$$\begin{array}{r} 40 \\ +26 \\ \hline 60 \end{array} \qquad \begin{array}{r} 31 \\ +18 \\ \hline 38 \end{array} \qquad \begin{array}{r} 70 \\ +15 \\ \hline 70 \end{array}$$

Description of Pattern _____

_____

PAPER 3

```
  67        48        5 3
+ 25      + 37      + 4 9
  82        75        9 2
```

Description of Pattern _____

_____

PAPER 4

```
   1         1 1        1 1
  28        2 4 8      4 5 7
  29          6 8      3 6 8
+ 34      + 1 6 5    + 1 9 2
  82        4 7 2      9 2 7
```

Description of Pattern _____

_____

PAPER 5

```
  6 4 5      4 8 2      5 7 6
+ 2 3 7    + 3 6 3    + 1 8 9
  8 7 1      7 1 5      6 1 1
```

Description of Pattern _____

_____

PAPER 6

```
    2          7          9
    6          5          8
  + 3        + 6        + 7
   29         8 1      1 0 5
```

Description of Pattern _____

_____

Paper 7

```
  1            1            1
  1 5          8 4          1 6
+ 6 7        + 5 6        + 8 1
─────        ─────        ─────
  8 2          1 0          1 3
```

Description of Pattern _____

_____

Paper 8

```
  1 6 5        3 6 9        4 5 4 3
+   3 3      +   5 6      +     6 9
───────      ───────      ─────────
  4 9 8        8 11 5      10 11 10 12
```

Description of Pattern _____

_____

Paper 9

```
  11                  11               11
   8 6 4           4 7 8            7 7 5
 + 5 8 9         + 2 9 5          + 4 8 3
 ───────         ───────          ───────
   3 3 4 3         2 6 6 3          2 1 5 8
```

Description of Pattern _____

_____

Paper 10

```
  4 7          6 5          7 8
-   3        -   2        -   4
─────        ─────        ─────
  1 4          4 3          3 4
```

Description of Pattern _____

_____

PAPER 11

$$\begin{array}{r} 6\,2 \\ -\ \ 5 \\ \hline 7\,5 \end{array} \qquad \begin{array}{r} 8\,4 \\ -\ \ 8 \\ \hline 6\,7 \end{array} \qquad \begin{array}{r} 5\,1 \\ -\ \ 3 \\ \hline 8\,4 \end{array}$$

Description of Pattern _____

_____

PAPER 12

$$\begin{array}{r} 6\,5 \\ -\,2\,9 \\ \hline 4\,6 \end{array} \qquad \begin{array}{r} 4\,3\,7 \\ -\ \ 8\,4 \\ \hline 4\,5\,3 \end{array} \qquad \begin{array}{r} 2\,2\,6 \\ -\,1\,7\,3 \\ \hline 1\,5\,3 \end{array}$$

Description of Pattern _____

_____

PAPER 13

$$\begin{array}{r} 2\,4\,8 \\ -\ \ 7\,5 \\ \hline 7\,3 \end{array} \qquad \begin{array}{r} 7\,3\,4 \\ -\ \ 6\,9 \\ \hline 6\,5 \end{array} \qquad \begin{array}{r} 5\,6\,5 \\ -\ \ 9\,8 \\ \hline 6\,7 \end{array}$$

Description of Pattern _____

_____

PAPER 14

$$\begin{array}{r} 5\,2 \\ -\,2\,7 \\ \hline 3\,0 \end{array} \qquad \begin{array}{r} 6\,1\,5 \\ -\,1\,4\,2 \\ \hline 5\,0\,3 \end{array} \qquad \begin{array}{r} 3\,2\,2 \\ -\,1\,5\,6 \\ \hline 2\,0\,0 \end{array}$$

Description of Pattern _____

_____

PAPER 15

$$\begin{array}{r} \overset{5}{}\overset{}{6}{}^{1}48 \\ -397 \\ \hline 851 \end{array} \qquad \begin{array}{r} 4\overset{2}{3}\overset{1}{6} \\ -218 \\ \hline 248 \end{array} \qquad \begin{array}{r} \overset{4}{5}{}^{1}29 \\ -385 \\ \hline 644 \end{array}$$

Description of Pattern _____

_____

PAPER 16

$$31 - 7 = \underline{22} \qquad\qquad 23 - 4 = \underline{13}$$
$$42 - 5 = \underline{33} \qquad\qquad 51 - 3 = \underline{46}$$

Description of Pattern _____

_____

PAPER 17

$$\begin{array}{r} 539 \\ -83 \\ \hline 206 \\ 351 \\ \hline 557 \end{array} \qquad \begin{array}{r} 457 \\ -65 \\ \hline 102 \\ 211 \\ \hline 313 \end{array} \qquad \begin{array}{r} 928 \\ -34 \\ \hline 524 \\ 615 \\ \hline 1139 \end{array}$$

Description of Pattern _____

_____

PAPER 18

$$\begin{array}{r} 705 \\ -108 \\ \hline 507 \end{array} \qquad \begin{array}{r} 602 \\ -238 \\ \hline 274 \end{array} \qquad \begin{array}{r} 304 \\ -176 \\ \hline 38 \end{array}$$

Description of Pattern _____

_____

PAPER 19

$$
\begin{array}{r}
9\,8 \\
\times\,1\,3 \\
\hline
2\,9\,4 \\
9\,8 \\
\hline
3\,9\,2
\end{array}
\qquad
\begin{array}{r}
3\,7 \\
\times\,2\,4 \\
\hline
1\,4\,8 \\
7\,4 \\
\hline
2\,2\,2
\end{array}
\qquad
\begin{array}{r}
5\,6 \\
\times\,3\,2 \\
\hline
1\,1\,2 \\
1\,6\,8 \\
\hline
2\,8\,0
\end{array}
$$

Description of Pattern _____

_____

PAPER 20

$$
\begin{array}{r}
7\,2\,3 \\
\times\,\quad 6 \\
\hline
4\,3\,3\,8
\end{array}
\qquad
\begin{array}{r}
3\,6\,8 \\
\times\,\quad 6 \\
\hline
1\,9\,7\,8
\end{array}
\qquad
\begin{array}{r}
4\,7\,5 \\
\times\,\quad 9 \\
\hline
3\,7\,4\,5
\end{array}
$$

Description of Pattern _____

_____

PAPER 21

$$
\begin{array}{r}
3\,6 \\
\times\,2\,5 \\
\hline
1\,8\,0 \\
1\,0\,2 \\
\hline
1\,2\,0\,0
\end{array}
\qquad
\begin{array}{r}
7\,8 \\
\times\,4\,3 \\
\hline
2\,3\,4 \\
3\,3\,2 \\
\hline
3\,5\,5\,4
\end{array}
\qquad
\begin{array}{r}
6\,5 \\
\times\,3\,7 \\
\hline
4\,5\,5 \\
2\,2\,5 \\
\hline
2\,7\,0\,5
\end{array}
$$

Description of Pattern _____

_____

PAPER 22

$$\overset{4}{3}\,7 \times 6 = 72$$

$$\overset{1}{8}\,5 \times 3 = 95$$

$$\overset{2}{2}\,5 \times 4 = 40$$

Description of Pattern _____

_____

PAPER 23

```
  4 3 6
X   2 5
  2 1 8 0
  8 7 2
  6 6 6 0
```

```
  3 7 9
X   4 2
    7 5 8
  1 5 1 6
  1 5 6 1 8
```

```
    7 5 4
  X 2 6 8
  6 0 3 2
  4 5 2 4
  1 5 0 8
  1 1 1 6 1 2
```

Description of Pattern _____

_____

PAPER 24

```
  1 3 2
X     6
  1 5 1 2
```

```
  3 5 8
X     4
  1 5 2 2
```

```
  4 9 2
X     7
  3 2 6 4
```

Description of Pattern _____

_____

PAPER 25

```
  4 1
X 3 6
  1 4
```

```
  5 2 6
X   2 5
  5 4 3
```

```
  8 3 7
X 2 9 4
  1 2 2
```

Description of Pattern _____

_____

PAPER 26

$$
\begin{array}{r}
{}^{42}_{}{}^{1} \\
48 \\
\times\ 63 \\
\hline
28944
\end{array}
\qquad
\begin{array}{r}
{}^{3}_{}{}^{3} \\
67 \\
\times\ 35 \\
\hline
20435
\end{array}
\qquad
\begin{array}{r}
{}^{6}_{}{}^{7} \\
92 \\
\times\ 48 \\
\hline
37536
\end{array}
$$

Description of Pattern _____

_____

PAPER 27

$$
\begin{array}{r}
5402 \\
\times\qquad 6 \\
\hline
32502
\end{array}
\qquad
\begin{array}{r}
603 \\
\times\quad 27 \\
\hline
4401 \\
1206 \\
\hline
16461
\end{array}
\qquad
\begin{array}{r}
8704 \\
\times\quad 74 \\
\hline
34906 \\
61108 \\
\hline
645986
\end{array}
$$

Description of Pattern _____

_____

PAPER 28

$$
\begin{array}{r}
72 \\
\times\ 43 \\
\hline
216 \\
2892 \\
\hline
3108
\end{array}
\qquad
\begin{array}{r}
43 \\
\times\ 23 \\
\hline
129 \\
866 \\
\hline
995
\end{array}
\qquad
\begin{array}{r}
52 \\
\times\ 14 \\
\hline
208 \\
524 \\
\hline
732
\end{array}
$$

Description of Pattern _____

_____

PAPER 29

$$
\begin{array}{r}
57 \\
\times\,34 \\
\hline
228 \\
1710 \\
\hline
17328
\end{array}
\qquad
\begin{array}{r}
96 \\
\times\,42 \\
\hline
192 \\
3840 \\
\hline
38592
\end{array}
\qquad
\begin{array}{r}
175 \\
\times\,53 \\
\hline
525 \\
8750 \\
\hline
88025
\end{array}
$$

Description of Pattern _____

_____

PAPER 30

$$
\begin{array}{r}
252\,R3 \\
4\,\overline{)129} \\
8 \\
\hline
21 \\
20 \\
\hline
11 \\
8 \\
\hline
3
\end{array}
\qquad
\begin{array}{r}
357 \\
4\,\overline{)1230} \\
28 \\
\hline
22 \\
20 \\
\hline
12 \\
12 \\
\hline
\end{array}
\qquad
\begin{array}{r}
514\;R3 \\
6\,\overline{)3924} \\
24 \\
\hline
9 \\
6 \\
\hline
33 \\
30 \\
\hline
3
\end{array}
$$

Description of Pattern _____

_____

PAPER 31

$$
\frac{20}{4} = 5 \qquad \frac{4}{12} = 3 \qquad \frac{7}{30} = 4\frac{2}{7}
$$

Description of Pattern _____

_____

PAPER 32

$$\frac{2}{3} + \frac{1}{4} = 37 \qquad \frac{5}{6} + \frac{1}{2} = 68 \qquad \frac{1}{5} + \frac{3}{4} = 49$$

Description of Pattern _____

_____

PAPER 33

$$\frac{1}{3} + \frac{2}{9} = \frac{3}{9} \qquad \frac{3}{4} + \frac{3}{2} = \frac{6}{4} \qquad \frac{5}{6} + \frac{1}{2} = \frac{6}{6}$$

Description of Pattern _____

_____

PAPER 34

$$\frac{3}{4} + \frac{1}{2} = 55 \qquad \frac{2}{3} + \frac{2}{5} = 75 \qquad \frac{5}{6} + \frac{1}{3} = 87$$

Description of Pattern _____

_____

PAPER 35

$$\frac{1}{6} + \frac{2}{3} = 12 \qquad \frac{3}{4} + \frac{1}{5} = 13 \qquad \frac{7}{8} + \frac{2}{6} = 23$$

Description of Pattern _____

_____

PAPER 36

$$
\begin{array}{l}
6\frac{1}{2} = \frac{2}{4} \\
+\,7\frac{1}{4} = \frac{1}{4} \\
\hline
\phantom{+7\frac{1}{4}=}\frac{3}{4}
\end{array}
\qquad
\begin{array}{l}
10\frac{5}{6} = \frac{5}{6} \\
+25\frac{2}{3} = \frac{4}{6} \\
\hline
\phantom{+25\frac{2}{3}=}\frac{9}{6} = 1\frac{1}{2}
\end{array}
\qquad
\begin{array}{l}
24\frac{1}{2} = \frac{4}{8} \\
+17\frac{5}{8} = \frac{5}{8} \\
\hline
\phantom{+17\frac{5}{8}=}\frac{9}{8} = 1\frac{1}{8}
\end{array}
$$

Description of Pattern _____

_____

PAPER 37

$$1\frac{3}{5} = 1\frac{8}{15}$$
$$+\, 2\frac{1}{3} = 2\frac{7}{15}$$
$$3\frac{15}{15}$$

$$4\frac{2}{4} = 4\frac{18}{4}$$
$$+\, 1\frac{1}{2} = 1\frac{3}{4}$$
$$5\frac{21}{4}$$

$$6\frac{2}{3} = 6\frac{20}{6}$$
$$+\, 3\frac{1}{6} = 3\frac{19}{6}$$
$$9\frac{39}{6}$$

Description of Pattern _____

_____

PAPER 38

$$\frac{4}{6} - \frac{1}{3} = \frac{3}{3} \qquad \frac{3}{4} - \frac{1}{2} = \frac{2}{2} \qquad \frac{5}{8} - \frac{2}{4} = \frac{3}{4}$$

Description of Pattern _____

_____

PAPER 39

$$\frac{5}{6} - \frac{4}{5} = \frac{1}{30} \qquad \frac{4}{5} - \frac{1}{2} = \frac{3}{10} \qquad \frac{5}{6} - \frac{3}{5} = \frac{2}{30}$$

Description of Pattern _____

_____

PAPER 40

$$3\frac{5}{12}$$
$$-1\frac{5}{12}$$
$$20$$

$$5\frac{2}{3}$$
$$-2\frac{1}{3}$$
$$3\frac{1}{3}$$

$$9\frac{3}{4}$$
$$-4\frac{3}{4}$$
$$50$$

Description of Pattern _____

_____

PAPER 41

$$5\frac{3}{4} = 5\frac{3}{4}$$
$$-2\frac{1}{2} = 2\frac{2}{4}$$
$$\overline{\phantom{-2\frac{1}{2} =}\; 3\frac{1}{4}}$$

$$7\frac{1}{6} = \cancel{7}^{6}\frac{\cancel{\times}^{6}}{6}$$
$$-6\frac{2}{3} = 6\frac{4}{6}$$
$$\overline{\phantom{-6\frac{2}{3} =}\; \frac{2}{6} = \frac{1}{3}}$$

$$8\frac{1}{3} = \cancel{8}^{7}\frac{\cancel{\times}^{6}}{6}$$
$$-2\frac{1}{2} = 2\frac{3}{6}$$
$$\overline{\phantom{-2\frac{1}{2} =}\; 5\frac{3}{6} = 5\frac{1}{2}}$$

Description of Pattern _____

_____

PAPER 42

$$7 \qquad = 6\frac{7}{7} = 6\frac{28}{28}$$
$$-3\frac{3}{4} = 3\frac{3}{4} = 3\frac{21}{28}$$
$$\overline{\phantom{-3\frac{3}{4} =}\quad 3\frac{7}{28} = 3\frac{1}{4}}$$

$$5 \qquad = 4\frac{5}{5} = 4\frac{15}{15}$$
$$-2\frac{1}{3} = 2\frac{1}{3} = 2\frac{5}{15}$$
$$\overline{\phantom{-2\frac{1}{3} =}\quad 2\frac{10}{15} = 2\frac{2}{3}}$$

$$9 \qquad = 8\frac{9}{9} = 8\frac{45}{45}$$
$$-1\frac{4}{5} = 1\frac{4}{5} = 1\frac{36}{45}$$
$$\overline{\phantom{-1\frac{4}{5} =}\quad 7\frac{9}{45} = 7\frac{1}{5}}$$

Description of Pattern _____

_____

PAPER 43

$$\frac{7}{8} \times \frac{3}{7} = \frac{7}{8} \times \frac{6}{8} = \frac{42}{8} \qquad \frac{1}{6} \times \frac{2}{3} = \frac{1}{6} \times \frac{4}{6} = \frac{4}{6}$$

$$\frac{5}{12} \times \frac{1}{4} = \frac{5}{12} \times \frac{3}{12} = \frac{15}{12}$$

Description of Pattern _____

_____

Paper 44

$$4\frac{2}{3} \times \frac{1}{4} = 4\frac{2}{12} \qquad 6\frac{3}{4} \times \frac{2}{3} = 6\frac{6}{12}$$

$$9\frac{1}{2} \times \frac{3}{4} = 9\frac{3}{8}$$

Description of Pattern _____

_____

Paper 45

$$5\frac{1}{3} \times 6\frac{3}{4} = 30\frac{3}{12}$$

$$7\frac{2}{5} \times 2\frac{1}{8} = 14\frac{2}{40} \qquad 1\frac{7}{8} \times 4\frac{2}{3} = 4\frac{14}{24}$$

Description of Pattern _____

_____

Paper 46

$$2\frac{1}{2} \times 1\frac{3}{10}$$
$$2\frac{5}{10} \times 1\frac{13}{10} = 2\frac{65}{10}$$

$$1\frac{3}{4} \times 3\frac{1}{8} \qquad\qquad\qquad 2\frac{2}{3} \times 2\frac{1}{6}$$
$$1\frac{7}{8} \times 3\frac{25}{8} = 3\frac{175}{8} \qquad 2\frac{8}{6} \times 2\frac{13}{6} = 4\frac{104}{6}$$

Description of Pattern _____

_____

PAPER 47

$$\frac{4}{8} \div \frac{2}{8} = \frac{8}{8}$$

$$\frac{5}{10} \div \frac{3}{10} = \frac{15}{10} \qquad\qquad \frac{3}{4} \div \frac{1}{2} = \frac{3}{4} \div \frac{2}{4} = \frac{6}{4}$$

Description of Pattern _____

_____

PAPER 48

$$\frac{5}{6} \div \frac{2}{3} = \frac{5}{6} \div \frac{4}{6} = \frac{1}{6}$$

$$\frac{1}{2} \div \frac{3}{4} = \frac{2}{4} \div \frac{3}{4} = \frac{1}{4} \qquad \frac{7}{8} \div \frac{1}{5} = \frac{35}{40} \div \frac{8}{40} = \frac{4}{40}$$

Description of Pattern _____

_____

PAPER 49

$$\frac{7}{8} \div \frac{1}{2} = \frac{7}{8} \div \frac{4}{8} = 1\frac{3}{8} \qquad\qquad \frac{8}{9} \div \frac{2}{3} = \frac{8}{9} \div \frac{6}{9} = 1\frac{2}{9}$$

$$\frac{11}{12} \div \frac{3}{4} = \frac{11}{12} \div \frac{9}{12} = 1\frac{2}{12}$$

Description of Pattern _____

_____

PAPER 50

$$.4 + .3 = \underline{.07} \qquad \begin{array}{r} 1.32 \\ +3.46 \\ \hline .04\ 78 \end{array} \qquad \begin{array}{r} 27.5 \\ +\ 8.9 \\ \hline 3.64 \end{array}$$

Description of Pattern _____

_____

(This pattern is commonly seen in seventh-grade classrooms.)

Paper 51

$$7.7 + 13.2 = ?\qquad 2.5 + 4.32 = ?\qquad 15.4 + 8.69 = ?$$

$$
\begin{array}{r}
7.7\\
+13.2\\
\hline
20.9
\end{array}
\qquad
\begin{array}{r}
2.\;5\\
+4.32\\
\hline
6.37
\end{array}
\qquad
\begin{array}{r}
15.\;4\\
+\;\;8.69\\
\hline
23.73
\end{array}
$$

Description of Pattern _____

_____

Paper 52

$$
\begin{array}{r}
63.5\\
\times\;\;0.7\\
\hline
444.5
\end{array}
\qquad
\begin{array}{r}
7.36\\
\times\;2.4\\
\hline
294.4\\
1472.0\\
\hline
1766.4
\end{array}
\qquad
\begin{array}{r}
45.82\\
\times\quad5.6\\
\hline
27492\\
229100.0\\
\hline
25659.2
\end{array}
$$

Description of Pattern _____

_____

PAPER 53

$$0.5\overline{)60} \rightarrow 0.5\overline{)\begin{array}{r} 12 \\ 60 \\ \underline{5} \\ 10 \\ \underline{10} \end{array}}$$

$$60\overline{)2.4} \rightarrow 60\overline{)\begin{array}{r} 0.04 \\ 2.40 \\ \underline{2\,40} \end{array}}$$

$$0.4\overline{)18} \rightarrow 0.4\overline{)\begin{array}{r} 4.5 \\ 18.0 \\ \underline{16} \\ 20 \\ \underline{20} \end{array}}$$

Description of Pattern _____

_____

PAPER 54

$$.7 = \frac{1}{10} \qquad .25 = \frac{7}{100} \qquad .302 = \frac{5}{1000}$$

Description of Pattern _____

_____

Paper 55

25% of 100 $\underline{2500}$

½ of 20% of 100 $\underline{1000}$

⅓ of 1% of 40 $\underline{8}$

Description of Pattern _____

_____

Key to Error Patterns in Additional Children's Papers

1. The sum of all digits is determined, regardless of place value.
2. The child is applying to addition the properties of zero and one for multiplication.
3. This child happens to be adding from left to right. When he adds the ones he does not know what to do with the tens digit, so he ignores it. If you merely look at his paper it is clear that the tens digit is being ignored; but you have to watch him compute or interview him to learn that he is adding left to right.
4. The child writes the greater digit and "carries" the lesser digit to the next column.
5. When the sum is ten or more, the child writes the tens digit and ignores the units digit.
6. The top two numbers may be perceived as a two-digit number, and the third number added or counted on (e.g., 2 + 6 + 3 is 26 + 3). Rodney Shover noticed that it is also possible the child is adding from the bottom up. For the first pair of addends, the units digit is recorded in the answer; if there is a ten it is added to the top digit to make the tens number in the sum.
7. When a sum in the tens column has two digits, write only the "tens" digit (actually, the digit is the number of hundreds.) "There is no room for the other number."
8. Sums are written fully (place values are ignored) and the left digit in the shorter numeral is added repeatedly.
9. All ones which result from regrouping are collected above the left-hand column. The number of ones is then written in the thousands place.
10. The minuend is subtracted from both the ones and the tens.
11. The child counts backwards to determine the missing addend but reverses the digits when recording the number.

12. The child is not remembering to subtract one ten (or one hundred) when she regroups.

13. Though regrouping correctly, the child does not subtract hundreds if there are no hundreds in the subtrahend.

14. The child thinks, "seven from two is nothing;" or he just writes zero when he "can't subtract."

15. After renaming, both hundreds digits are *added* before subtracting; a similar procedure is used with tens.

16. The minuend is rounded down to the nearest decade, then the known addend is subtracted. Finally, the number of units in the minuend is subtracted.

17. Using the computational sequence associated with multiplication, the child is comparing digits and recording each difference (S-W-1). As is true for multiplication, addition precedes the final answer.

18. The child regroups directly from hundreds to ones.

19. The second partial product is placed incorrectly.

20. In regrouping, one is always added regardless of the number required.

21. When multiplying by the tens digit, both crutch figures are used.

22. For the tens, the child just adds without multiplying first.

23. The partial products are "subtracted" by finding differences.

24. After adding the number of tens to be remembered, the *tens* digit is recorded (in the tens place!) and the units digit is "carried."

25. Basically, an addition procedure is used with multiplication facts (M-W-4), but when the product for a fact has two digits the tens figure is recorded.

26. Multiplication facts are recalled in the correct sequence, but crutches are recorded at the left, then the right, and then the left again. Place values are ignored in the product.

27. When there is a zero in the minuend, the child inserts a zero before regrouping.

28. The tens digit is used to multiply the other three digits, proceeding counterclockwise. Also, the second partial product is placed incorrectly.

29. For the second partial product, this student moves over one place *and also* writes a zero. It is likely that both procedures were learned as rote procedures and the child does not realize that either procedure by itself is sufficient to multiply by ten as needed.

30. Division proceeds from right to left. As one child using this procedure said, "We always start with the ones." Place values are necessarily confused.

31. The greater number (numerator *or* denominator) is divided by the lesser.

32. The numerators are added and recorded (as tens), then the denominators are added and recorded to the right (as ones).

33. The numerators are added and recorded as the new numerator. The greater denominator is used as the new denominator because the lesser denominator "will go into" the greater denominator.

34. The first denominator and second numerator are added and written as the number of units; then the first numerator and second denominator are added for the tens digit.

35. All numerators and denominators are added together as if they were whole numbers, though the explanation may sound something like: "2 over 3 is 5, 1 over 6 is 7, and 5 plus 7 is 12."

36. This student apparently becomes so involved with the process of renaming fractions that she forgets to add the whole numbers.

37. After the least common denominator is determined, the child computes each numerator by multiplying the original denominator times the whole number and then adding the original numerator.

38. This student subtracted the numerators and also subtracted the denominators.

39. The numerators are subtracted and the denominators are multiplied. Why does this work for two of the three examples? For help see Mona S. Haddad, "An Error Pattern Leads to a Discovery Lesson," in *The Mathematics Teacher* 73, no. 3 (March 1980): 197.

40. When the difference between the fractions is zero, the zero is recorded as a whole number in what becomes the one's place—effectively multiplying the answer by ten.

41. When renaming a mixed number so that the fraction in the minuend will be as great or greater than the fraction in the subtrahend, this student subtracts from the whole number without properly adding an equivalent amount to the fraction part of the mixed number. In the mind of the student, it may be that the one that was subtracted is shown with the new fraction in the form n/n.

42. The whole number is always used for the numerator and denominator when creating n/n, even though this unnecessarily complicates the procedure with unlike denominators. You may want to call this an inefficient algorithm rather than an error pattern, for it *does* produce the correct answer.

43. This student renames unlike fractions so they have a common denominator before multiplying the numerators. The common denominator is used for the denominator in the product. The student's procedure is very similar to the correct procedure for *adding* unlike fractions.

44. The common fractions are multiplied, with the result affixed to the whole number.

45. The whole numbers and common fractions are multiplied independently.

46. After a common denominator is determined, a new numerator is computed by multiplying the original denominator times the whole number and then adding the original numerator. The whole numbers are multiplied; only the numerators are multiplied in the common fractions. This is an example of how an error pattern adopted for one operation (see Paper 35) is used to create other erroneous procedures.

47. Common denominators, if needed, are provided first; then the numerators are multiplied to determine the resulting numerator.

48. After common denominators are determined, one numerator is divided by the other (with remainder ignored) to determine the resulting numerator.

49. The divisor is not inverted. Instead, this student finds the equivalent fraction with the same denominator as the dividend, then proceeds much as with adding unlike fractions. The numerators are divided as if they were whole numbers, and the remainder is placed over the common denominator for the fraction part of the mixed-number answer.

50. The rule for placing the decimal point in a product is applied to a sum.

51. The student is careful to do two things he has been taught: line up the digits on the right, and line up the decimal points.

52. This student is overgeneralizing from addition computation. The decimal point in the product is simply aligned with the decimal point in the bottom factor.

53. This student copies the example, but without any decimal points, and then divides as if both numbers were whole numbers. After dividing, decimal points are placed within the dividend and the divisor as they were originally. Finally, a point is placed within the quotient above the decimal point in the dividend.

54. The denominator is determined by the number of places in the decimal; digits are added to find the numerator. This misunderstanding is sometimes prompted by the explanation:

$$.25 = \frac{2}{10} + \frac{5}{100}$$

55. This student multiplies all of the top numbers as if they were whole numbers, then divides by the "bottom number" (denominator).

APPENDIX
B

Error Patterns in Other Areas of Mathematics

The phenomenon of students adopting erroneous procedures is not limited to the simpler computations of arithmetic, as is clearly illustrated in the examples which follow. Look carefully at each student's paper. Can you determine the error pattern? Why might a student learn such procedures? Are there things you can do as you teach mathematics that will make it less likely that students adopt erroneous patterns?

PAPER A

1. 57 = _5_ tens + _7_ ones

2. 483 = _4_ ones + _8_ hundreds + _3_ tens

3. 270 = _2_ hundreds + _7_ ones + _0_ tens

PAPER B

1. ☐ − 384 = 126   ☐ = _258_

2. ☐ × 13 = 260   ☐ = _3380_

3. ☐ ÷ 40 = 20   ☐ = _2_

Paper C

1. 65 ÷ 13 = □	□ = _5_
2. □ ÷ 12 = 36	□ = _432_
3. 17 × □ = 68	□ = _4_
4. 60 ÷ □ = 30	□ = _1800_
5. 24 × 8 = □	□ = _192_
6. 90 ÷ □ = 15	□ = _1350_

Although the unknown was correctly identified in four of the six examples, the same error pattern was applied to all six. For an interesting discussion of this pattern, see Barbara R. Sadowski and Delayne H. McIlveen, "Diagnosis and Remediation of Sentence-solving Error Patterns," in *The Arithmetic Teacher* 31, no. 5 (January 1984): 42–45.

Paper D

1. John spent $4.50 at the fair. Now he has $2.75 remaining. How much money did he have before the fair?

$1.75

2. The store had a sale of red and blue shirts. There were 46 red shirts left after the sale, and 28 blue shirts were left. How many shirts were left after the sale?

18

Are "key words" effective in problem solving?

Paper E

1. If a 9-pound ham costs $17.01, what is the cost of a 12-pound ham?

$22.68

2. Mrs. Jones gave $10 to Mark and asked him to put 5 gallons of gasoline in her car. Gasoline sells for $1.23 a gallon. What is her change?

$48.77

The order in which operations are presented is not always the order in which they are computed. Reordering data is particularly difficult for the child who tends to use each number in the sequence presented.

Paper F

> Find the average for each set of numbers.
>
> A. 26, 74, 83   Answer___45___
>
> B. 73, 98, 65   Answer___59___
>
> C. 57, 62, 95, 81   Answer___59___

This child probably records a crutch when adding.

Paper G

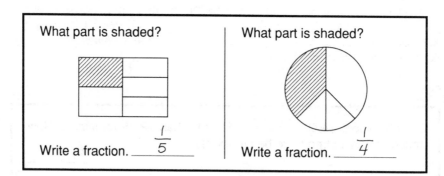

> What part is shaded?
>
> Write a fraction. $\frac{1}{5}$

> What part is shaded?
>
> Write a fraction. $\frac{1}{4}$

Paper H

> 1. $^-8 + 6 =$ ___$^-2$___   3. $7 + ^-2 =$ ___5___
>
> 2. $5 + ^-9 =$ ___4___   4. $^-4 + 10 =$ ___-6___

This child probably made his rule while working with examples like 1 and 3 in which the sign of the first addend is also the sign of the sum.

Paper I

> Write the sum.
>
> 1. $^-6 + ^-8 =$ ___2___   2. $4 + ^-3 =$ ___7___
>
> 3. $^-8 + ^-3 =$ ___$^-5$___   4. $^-2 + ^-5 =$ ___3___

Why does this child use the sign for one number if she ignores the sign for the other number?

PAPER J

1. $3 - (^-4) = $ __7__	3. $7 + ^-2 = $ __5__
2. $^-6 - 2 = $ __8__	4. $^-5 - 4 = $ __9__

Do two negatives make a positive?

PAPER K

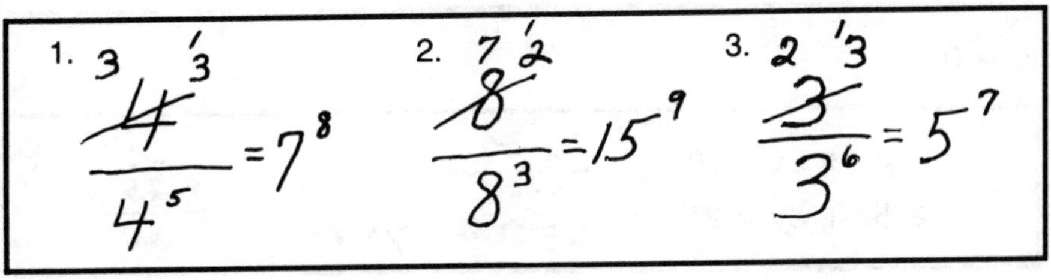

1. $\dfrac{3^{\prime}3}{\cancel{4}} = 7^8$ 4^5

2. $\dfrac{7^{\prime}2}{\cancel{8}} = 15^9$ 8^3

3. $\dfrac{2^{\prime}3}{\cancel{3}} = 5^7$ 3^6

PAPER L

$$1.\ 3(x + 2) = \underline{3x + 2}$$
$$*2.\ a(b + c) = \underline{ab + c}$$

Does "multiply" mean "remove the brackets"? (*One reviewer said with reference to this one: "I see it in calculus all of the time.")

PAPER M

$$1.\ 6(1 + 4x) + 2 = 6(5x) + 2$$
$$= 30x + 2$$
$$2.\ 7 + 5(2 + 3x) = 7 + 5(5x)$$
$$= 7 + 25x$$

Paper N

Add:

1. $\dfrac{1}{3b} + \dfrac{5a}{6c} = \dfrac{6a}{9bc}$

2. $\dfrac{3x}{2y} + \dfrac{4}{5z} = \dfrac{7x}{7yz}$

Paper O

1. $3x + 2x = 60$ $x = 5$ $\begin{array}{r} 35 \\ + 25 \\ \hline 60 \end{array}$

2. $5x + 3y = 81$ $x = 0$ $y = 1$

Paper P

1. $a^2 \cdot a^3 = \underline{\ a^6\ }$ 2. $y^4 \cdot y^3 = \underline{\ y^{12}\ }$

Paper Q

a. $2^3 \cdot 2^2 = \underline{\ 4^5\ }$ b. $3^2 \cdot 3^4 = \underline{\ 9^8\ }$

Paper R

1. $(a^2)^2 = \underline{\ a^4\ }$
2. $(b^2)^3 = \underline{\ b^5\ }$
3. $(a^3b^4) = \underline{\ a^5\,b^6\ }$

Paper S

1. $\dfrac{x + 3}{4 + x} = \dfrac{3}{4}$ 2. $\dfrac{8 + x}{x + 2} = 4$

What does it mean to "cancel like terms"?

P<small>APER</small> *T*

a. $\sqrt{3} + \sqrt{20} = \underline{\sqrt{23}}$ b. $\sqrt{5} + \sqrt{8} = \underline{\sqrt{13}}$

P<small>APER</small> *U*

a. $\dfrac{y + 2}{y + 4} = \dfrac{6}{9}$

Answer:

$\underline{y = 4\ or\ 5}$

b. $\dfrac{y - 3}{y + 1} = \dfrac{5}{9}$

Answer:

$\underline{y = 8}$

c. $\dfrac{y + 1}{y + 5} = \dfrac{7}{8}$

Answer:

$\underline{y = 6\ or\ 3}$

d. $\dfrac{y + 1}{y - 2} = \dfrac{3}{4}$

Answer:

$\underline{y = 2\ or\ 6}$

P<small>APER</small> *V*

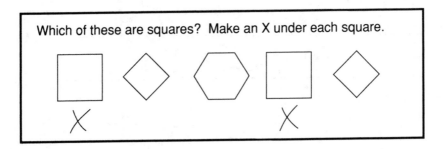

Which of these are squares? Make an X under each square.

This child's understanding of "square" includes a specific orientation. The child says, "If you turn a square and make a diamond, it's not a square anymore."

PAPER W

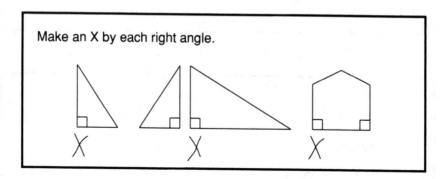

Make an X by each right angle.

Are some angles "left angles?"

PAPER X

1. The two pentagons are similar. Find the length of side X.

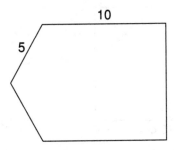

Answer _____ 6 _____

2. The two triangles are similar. Find the measure of angle Y.

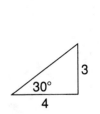

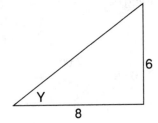

Answer _____ 60° _____

3. The two trapezoids are similar. Find the measure of angle Z.

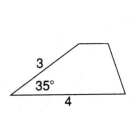

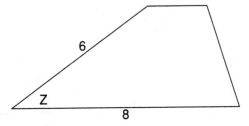

Answer _____ 70° _____

Paper Y

1. The two rectangles are similar. Find the length of side X.

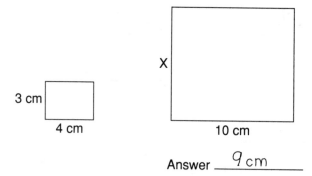

3 cm

4 cm

X

10 cm

Answer ___9 cm___

2. The two triangles are similar. Find the hypotenuse of the larger triangle.

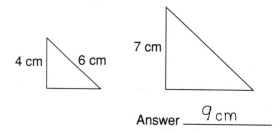

4 cm   6 cm

7 cm

Answer ___9 cm___

3. The two rectangles are similar. Find the diagonal of the smaller rectangle.

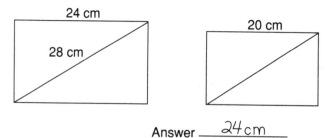

24 cm

28 cm

20 cm

Answer ___24 cm___

PAPER Z

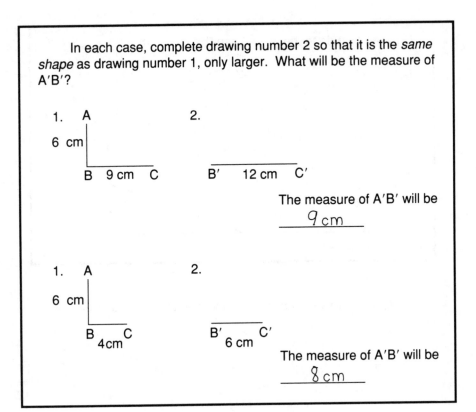

In each case, complete drawing number 2 so that it is the *same shape* as drawing number 1, only larger. What will be the measure of A'B'?

1. A

6 cm

B   9 cm   C

2.

B'   12 cm   C'

The measure of A'B' will be

9 cm

1. A

6 cm

B   C
  4cm

2.

B'   C'
  6 cm

The measure of A'B' will be

8 cm

*P*APER *AA*

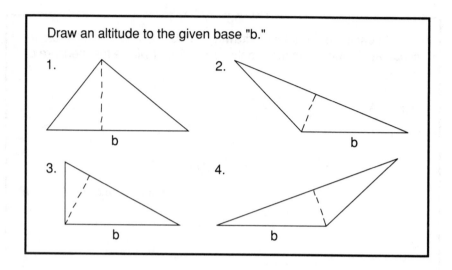

What assumption is this student making about an altitude to a given side?

Paper BB

Determine the perimeter of each of these rectangular regions:

1.

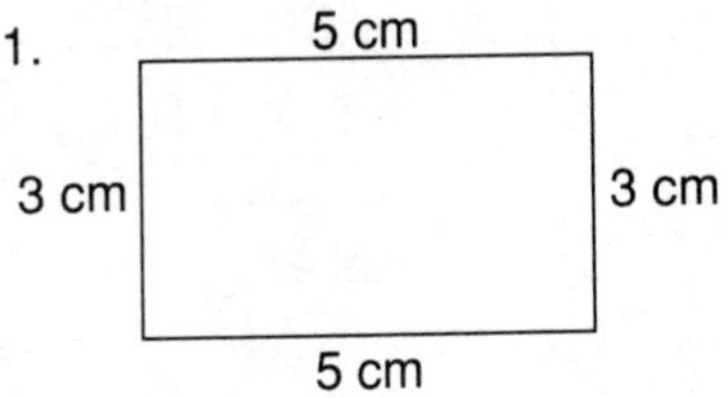

5 cm

3 cm   3 cm

5 cm

Answer ___16cm___

2.

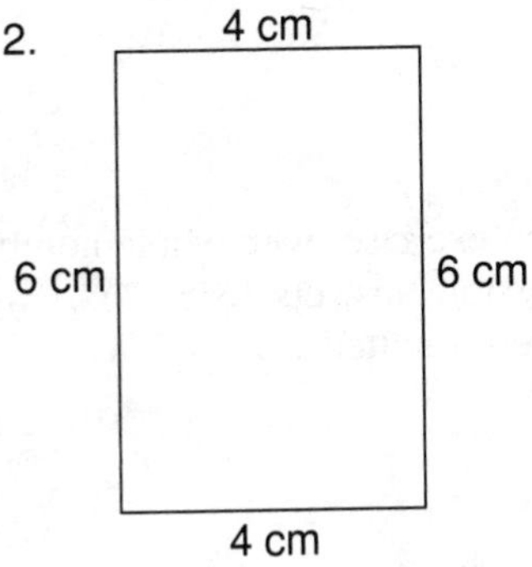

4 cm

6 cm   6 cm

4 cm

Answer ___20 cm___

3.

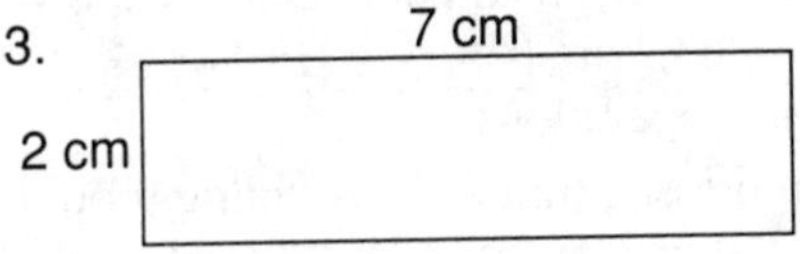

7 cm

2 cm

Answer ___14cm___

4.

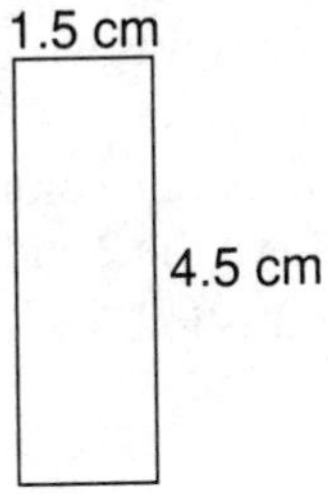

1.5 cm

4.5 cm

Answer ___6.75cm___

The Nature of
Wrong Answers[*]

WHOLE NUMBERS

There are 2173 possible answers to the thirteen exercises with whole numbers by the 176 pupils interviewed. . . . Of these possible answers 1658 (76%) were right; 449 (21%) were wrong; and 66 (3%) were omitted. . . .

Addition

1. Many combinations were recalled incorrectly in addition, as in "9 + 8 = 18, + 1 = 19 and 8 is 27," or "7 × 8 = 63," or "8 × 5 = 35," or "7 × 8 = 48." The same was true in each of the operations.
2. When counting was used pupils often lost count as in counting 9 on to 17 in 17 + 9 and getting 25, or in "7 × 8 = 57 (7 × 5 = 35; 7 × 6 = 42; 7 × 7 = 49; 50, 51, 52, 53, 54, 55, 56, 57)."
3. Many pupils failed to add the carried digit even when it was written above the top digit in the column to the left.
4. Sometimes the wrong digit was carried from the sum of one column to the next as in "2 + 3 = 5, + 9 = 14, + 7 = 21, put down the 2 and carry 1."
5. Intending to add, a pupil may have, in fact, multiplied as in "7 + 2 is 14, + 5 is 19, + 2 is 21, + 4 is 25."

Subtraction

6. Some pupils intended to subtract but in fact divided as in 93 − 32, "2 from 3 is 1; 3 from 9 is 3."

[*] From Francis G. Lankford, Jr., *Some Computational Strategies of Seventh Grade Pupils,* U. S. Department of Health, Education, and Welfare, Office of Education, National Center for Educational Research and Development (Regional Research Program) and The Center for Advanced Study, The University of Virginia, October 1972 (Project number 2-C-013, Grant number OEG-3-72-0035), pp. 27–33.

7. A wrong order was often used in subtraction as in 86 − 49, a pupil would say "9 minus 6 is 3; 8 − 4 is 4" or "9 from 6 leaves 3 and 4 from 8 leaves 4" or in 708 − 329 a pupil said "9 from 8 is 1; 2 from 0 is 0; 7 from 3 is 4."

8. A pupil would think to borrow to increase a digit but not reduce the digit from which borrowed, as in 86 − 49 "16 − 9 = 7 [counting]; 8 − 4 = 4."

9. When borrowing, as in 708 − 329, a pupil might borrow twice, once to make 0 a 10, and again to make 8 an 18, leaving the 7 as 5.

10. Some pupils borrowed from the tens column only, when they should have borrowed from both tens and hundreds columns, as in 708 − 329, rewrote 708 as 7−9−18, then "18 from 9 is 9; 9 from 2 is 7; 7 from 4 is 3."

11. Other pupils borrowed from the hundreds column only and rewrote the tens digit incorrectly. As in 708 − 329; rewritten as 6-10-18. Then 18 − 9 = 9; 10 − 2 = 8; 6 − 3 = 3.

12. The minuend was rewritten simply by affixing ones where needed, as in 708 − 329 which became 7-10-18 for 708 and the answer was 18 − 9 = 9; 10 − 2 = 8; 7 − 3 = 4.

Multiplication

13. The ones digit was multiplied by the ones digit and the tens digit by the tens digit only, as in 19 × 20 = _____, rewritten as 19, then "0 × 9 = 0 and 2 ×
                                                                           × 20

1 = 2" answer 20; or 58 "5 × 8 = 40; 7 × 5 = 35, + 4 = 39. Written as a
            × 75

single product 390.

14. The carried number was not included in the partial product, as in 19 × 20 = _____, rewritten as 19, then "0 × 9 = 0; 0 × 1 = 0; 2 × 9 = 18; 2 × 1 = 2."
                                            × 20

Throughout the remaining pages of this report a hyphen or hyphens after a numeral indicate an indentation in the arrangement of a partial product. For example, partial products 1824, 000, and 1520- were arranged by the pupil this way

$$
\begin{array}{r}
1824 \\
000 \\
1520 \\
\hline
17024
\end{array}
$$

or partial products 1824, 000-, 1520--were arranged this way

$$
\begin{array}{r}
1824 \\
000 \\
1520 \\
\hline
153824
\end{array}
$$

15. Place value of partial products was confused as in 19 "$0 \times 9 = 0; 0 \times 1 =$

$$\begin{array}{r} 19 \\ \times\,20 \\ \hline 3800 \end{array}$$

0; $2 \times 9 = 18; 2 \times 1 = 2, + 1 = 3$." Or in 304 "$6 \times 304 = 1824; 0 \times$

$$\begin{array}{r} 304 \\ \times\,506 \end{array}$$

$304 = 000; 5 \times 304 = 1520$-" for sum 17024.

16. The wrong product was written when one factor was 0, as in $19 \times 20 = $ ____; "$0 \times 9 = 9; 0 \times 1 = 1; 2 \times 9 = 18; 2 + 1 = 2, + 1 = 3$." Pupil wrote 38- under 19 for sum of 399.

17. A multiplication fact was recalled incorrectly as $7 \times 8 = 54$, in 58×75 "$8 \times 5 = 40; 5 \times 5 = 25, + 4 = 29; 7 \times 8 = 54; 7 \times 5 = 35, + 4 = 40$." Then $290 + 404- = 4330$.

18. One of the digits in the multiplier was not used in finding the product, as in

$$\begin{array}{r} 304 \\ \times\,506 \end{array}$$

only two partial products $6 \times 304 = 1824; 5 \times 304 = 1520$-. Sum 17024.

19. Partial products were found correctly but errors were made in adding them, as in 58 $5 \times 58 = 290; 7 \times 58 = 406$- for sum 4360, said "$9 + 6 = 16$" in
    $\times\,75$

    adding partial products.

Division

20. A remainder was interpreted wrong as in $27\overline{)81}$ $81 \div 27 = 3; 3 \times 27 = 81; 81 - 81 = 0$; "27 won't go into 0, so answer is 30"; or in $48\overline{)93}$ "48 goes into 93 one time; $1 \times 48 = 48; 93 - 48 = 45; 48$ can't go into 45; put 0 up; $45 - 0 = 45$" for answer 10 R45.

21. Long division was confused with short division as in $27\overline{)81}$ "2 goes into 8 four times; $2 \times 4 = 8$. Then $81 - 8- = 01; 2$ won't go into 1" so answer is 4 R1. Or in $48\overline{)93}$ "4 goes into 9 two times; $4 \times 2 = 8; 9 - 8 = 1$; bring down 3; 8 goes into 13 one time; $8 \times 1 = 8; 13 - 8 = 5$" answer 21 R5.

22. Quotient digit was multiplied by the divisor incorrectly, as in $74\overline{)6484}$ "8 × 74 = 572" ($8 \times 4 = 32; 8 \times 7 = 54, + 3 = 57$).

23. Errors were made in repeated multiplications to find quotient digit, as in $74\overline{)6484}$ decided 74 goes into 648 seven times, then $7 \times 74 = 658$ (thought $7 \times 4 = 28$ and $7 \times 7 = 56, + 7 = 63$).

24. Derived an answer before operation was complete, as in $74\overline{)6484}$ "74 goes into 684, eight times; $648 - 592 = 56$," so answer is 8 R56.

25. By repeated multiplication tried incorrectly to derive entire quotient instead of one digit at a time as in $74\overline{)6484}$; multiplied 74 by 12, by 24, by 52, and by 61. Chose 52 for quotient "because 3848 is closest to 6484"; then $6484 - 3848 = 2636$. Placed 2734 (incorrect product of 74×61) under 2636. Then $2636 - 2734 = 102$. Answer 5261 R102.

26. Place value was handled incorrectly in the quotient, as in 15)$\overline{7590}$ "15 goes into 75 five times 75 − 75 = 0; bring down your 9; 15 won't go into 9 so bring down 0; 6 × 15 = 90" so answer is 56. Or 15 into 75 five times; 75 − 75 = 0 "15 won't go into 0 so bring down 9; 15 won't go into 9 so bring down 0; 15 into 90 goes 6 times; 90 − 90 = 0; 15 into 0 zero times" so answer is 560.

FRACTIONS

There were 2640 possible answers to the sixteen exercises in computation with fractions by the 176 pupils interviewed. . . . Of these possible answers, 924 (35%) were right; 865 (33%) were wrong; and 851 (32%) were omitted. . . .

Addition

1. A prevalent practice was to add numerators and place the sum over one of the denominators or over a common denominator, as in $\frac{3}{4} + \frac{5}{2} = \frac{8}{4}$ "5 + 3 = 8. You don't add the bottom numbers because 2 will go into 4."
2. The most prevalent practice in adding fractions was to add numerators for the numerator of the sum and the same for the denominators, as in $\frac{3}{4} + \frac{5}{2} = \frac{8}{6}$ or $\frac{3}{8} + \frac{7}{8} = \frac{10}{16}$.
3. Many errors were made as pupils undertook to write equivalent fractions with common denominators, as in $\frac{3}{4} + \frac{5}{2} =$ _____; choose 4 as C.D. Then, for $\frac{3}{4}$, "4 times 1 equals 4 and 1 + 3 is 4," so $\frac{4}{4}$; for $\frac{5}{2}$ "2 × 2 = 4 and 4 × 5 = 20," so $\frac{20}{4}$; or $\frac{8}{8}$ for $\frac{7}{8}$ ("8 into 8 one time and 1 + 7 = 8"). The same thing was done in the other operations.
4. Several relatively large whole number answers were a surprise, as in $\frac{3}{4} + \frac{5}{2} = 86$ (5 + 3 = 8; 4 + 2 = 6) or $\frac{3}{4} + \frac{5}{2} = 59$ ("4 and 5 is 9; 3 and 2 is 5"), or $\frac{3}{8} + \frac{7}{8} = 26$ ("7 over 8 is 15; 3 over 8 is 11; 15 + 11 = 26").
5. The numerator and denominator of one fraction were added for the numerator of the sum, and the same with the second fraction for the denominator of the sum, as in $\frac{3}{8} + \frac{7}{8} = \frac{11}{15}$ ("8 and 3 is 11; 7 and 8 is 15"), or $\frac{3}{4} + \frac{5}{2} = \frac{7}{7}$ ("3 + 4 = 7; 2 + 5 = 7"), or $\frac{2}{3} + \frac{1}{2} = \frac{5}{3}$ ("2 + 3 = 5; 2 + 1 = 3").

Subtraction

6. As in addition, a very prevalent practice was to subtract numerators for the numerator of the difference and the same with denominators; as in $\frac{3}{4} - \frac{1}{2} = \frac{2}{3}$ (3 − 1 = 2; 4 − 2 = 2); or $8\frac{2}{5} - 4\frac{3}{10} = 4\frac{1}{5}$ (8 − 4 = 4; 3 − 2 = 1; 5 from 10 is 5), or $7\frac{1}{2} - 4\frac{1}{4} = 3\frac{0}{2}$ (7 − 4 = 3; 1 − 1 = 0; 2 from 4 = 2), or $\frac{5}{8} - \frac{1}{3} = \frac{4}{5}$ (1 from 5 is 4; 3 from 8 is 5).
7. In writing equivalent fractions, some pupils divided a denominator into the C.D. and added this quotient to the numerator of the original fraction for the

numerator of the equivalent fraction, as in $\frac{3}{4} = \frac{4}{4}$ (4 goes into 4 one time; $3 + 1 = 4$). Others subtracted for the new numerator, as in $\frac{5}{8} = \frac{2}{24}$ ("8 goes into 24, three times, 3 take away 5 is 2").

8. As in addition, some surprising whole numbers were derived for answers, as in $\frac{3}{4} - \frac{1}{2} = 22$ ("2 take away 4 is 2; 1 take away 3 is 2"), or $8\frac{2}{5} - 4\frac{3}{10} = 394$ ("2 over 5 would leave 3; 3 over 10 would leave 9; 4 from 8 would leave 4"), or $7\frac{1}{2} - 4\frac{1}{4} = 133$ ("1 over 2 leave 1; 1 over 4 would be 3; 7 from 4 would leave 3").

9. There were cases of the wrong use of borrowing, as in $8\frac{2}{5} - 4\frac{3}{10} = 3\frac{3}{5}$ (borrowed 1 from 8; made it a 7; changed 2 of $\frac{2}{5}$ into 12; then $7\frac{12}{5} - 4\frac{3}{10} = 3\frac{9}{5}$), or $8\frac{2}{5} - 4\frac{3}{10} = 3\frac{1}{10}$ (wrote $\frac{4}{10}$ for $\frac{2}{5}$ and $\frac{4}{10}$ for $\frac{3}{10}$; "you can't subtract 4 from 4, so you borrow 1 from 4 [remainder from $8 - 4$] make it a 3." Made first $\frac{4}{10}$ into $\frac{5}{10}$, then $\frac{5}{10} - \frac{4}{10} = \frac{1}{10}$).

10. A frequent error in writing equivalent fractions was to choose a C.D.; use it for the denominator of the new fraction but retain the numerator of the old fraction; as in $\frac{5}{8} = \frac{5}{24}$ and $\frac{1}{3} = \frac{1}{24}$.

11. The borrowed number was used incorrectly as in $9\frac{2}{3} - 5\frac{7}{8}$ rewritten as $9\frac{16}{24} - 5\frac{21}{24}$. Then $8\frac{26}{24} - 5\frac{21}{24}$.

Multiplication

12. Many pupils first wrote equivalent fractions, unnecessarily, and then incorrectly multiplied numerators and placed the product over the C.D., as in $\frac{2}{3} \times \frac{3}{5} = \frac{10}{15} \times \frac{9}{15} = \frac{90}{15}$, or $\frac{2}{3} = \frac{7}{15}$ ("3 goes into 15 five times; $5 + 2 = 7$") and $\frac{3}{5} = \frac{6}{15}$ ("5 goes into 15 three times; $3 + 3 = 6$"). Then $\frac{7}{15} \times \frac{6}{15} = \frac{46}{15}$ because $6 \times 7 = 46$, or $2\frac{1}{2} \times 6 = \frac{5}{2} \times \frac{12}{2} = \frac{60}{2}$.

13. Here, as in addition and subtraction, surprisingly large whole numbers were derived as products, as in $\frac{2}{3} \times \frac{3}{5} = 100$ ("$2 \times 5 = 10$, put down 0 and carry 1; $3 \times 3 = 9, + 1 = 10$. Answer 100"), or $2\frac{1}{2} \times 6 = 120$ (wrote vertically with 6 below $2\frac{1}{2}$. Then "0 times $\frac{1}{2} = 0$; there is nothing under $\frac{1}{2}$ so multiply by 0; $6 \times 2 = 12$, answer 120"), or $\frac{2}{3} \times \frac{3}{5} = 615$ ("$2 \times 3 = 6$; $3 \times 5 = 15$").

14. In all the operations there were examples of correctly derived answers with errors introduced with conversions to simpler form, as in $2\frac{1}{2} \times 6 = \frac{5}{2} \times \frac{6}{1} = \frac{30}{2} = 15\frac{1}{2}$ ("2 goes into 30 fifteen times, and the denominator is 2"), or $\frac{2}{3} \times \frac{3}{5} = \frac{2}{3}$ ("$\frac{2}{3} \times \frac{3}{5} = \frac{6}{15}$, to reduce divide by $\frac{3}{3}$; 6 goes into 3 two times; 15 goes into 3 three times, so that'll be $\frac{2}{3}$").

15. Some pupils wrote the reciprocal of the second factor before multiplying, as in $\frac{2}{3} \times \frac{3}{5} = \frac{2}{3} \times \frac{5}{3} = \frac{10}{9}$, or $2\frac{1}{2} \times 6 = \frac{5}{2} \times \frac{1}{6} = \frac{5}{12}$.

16. In a mixed number times a fraction the fractions would be multiplied and the whole number affixed, as in $5\frac{1}{2} \times \frac{3}{4} = 5\frac{3}{8}$ ("$1 \times 3 = 3$; $2 \times 4 = 8$; bring over 5"), or in $5\frac{1}{2} \times \frac{3}{4} = 5\frac{3}{2}$ ("$5\frac{1}{2} = 5\frac{2}{4}$, and $5\frac{2}{4} \times \frac{3}{4} = 5\frac{6}{4} = 5\frac{3}{2}$").

17. In a mixed number times a whole number the whole numbers would be multiplied and the fraction affixed, as in $2\frac{1}{2} \times 6 = 12\frac{1}{2}$ ("$6 \times 2 = 12$, bring over $\frac{1}{2}$").

Division

18. As in multiplication a widely used practice was to divide numerators and place the product [sic.] over the C.D., as in $9/10 \div 3/10 = 3/10$, or even in $7/8 \div 2/3 = 3/2$ ("2 goes into 7 three times; 3 goes into 8 two times"), or $15\frac{3}{4} \div \frac{3}{4} = 63/4 \div 3/4 = 21/4$.

19. After writing equivalent fractions, errors were made in dividing numerators, as in $7/8 \div 2/3 = 21/24 \div 16/24 = 15/24$, or $21/24 \div 16/24 = 1\ R5$.

20. In dividing a mixed number by a whole number, the whole number was divided by the whole number and the fraction was affixed, as in $69/10 \div 3 = 29/10$.

21. In dividing a mixed number by a fraction, the fractions were divided and the whole number affixed, as in $15\frac{3}{4} \div \frac{3}{4} = 15\frac{1}{4}$ ("$3 \div 3 = 1$, bring over 15, the answer is $15\frac{1}{4}$"), or $15\frac{3}{4} \div \frac{3}{4} = 16$ ("$\frac{3}{4} \div \frac{3}{4} = 1$, bring over 15 and $15 + 1 = 16$").

22. Numerators of like fractions were multiplied instead of divided, as in $9/10 \div 3/10 = 27/10$ ("the denominator would be 10; $3 \times 9 = 27$, and 27 would be numerator"), or in $15\frac{3}{4} \div \frac{3}{4} = 159/4$ ("bring over 15; $3 \times 3 = 9$; bring over 4").

23. Numerators and denominators were multiplied without writing a reciprocal of the divisor, as in $69/10 \div 3 = 69/10 \times 3/1 = 207/10$.

Sample Learning Hierarchy for Addition of Unlike Fractions

The hierarchy on the following page identifies specific skills which should be learned before other skills; simpler skills are at the bottom, with more complex tasks toward the top. When two or three skills are shown as immediately subordinate there is no specific order for teaching the subordinate skills, but *all* of the subordinate skills should be learned before the next higher skill. Note that multiplication of fractions precedes addition of unlike fractions.

Of course, specific skills in the hierarchy are taught over time. Many different instructional activities are typically required to teach each skill.

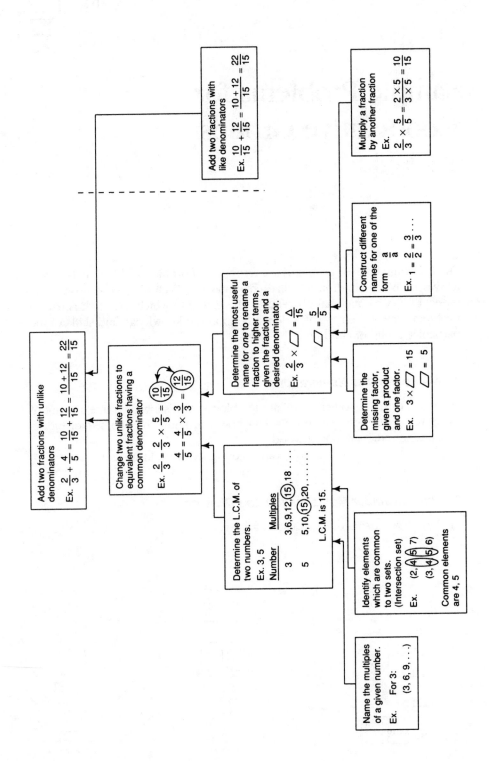

Add two fractions with like denominators
Ex. $\frac{10}{15} + \frac{12}{15} = \frac{10+12}{15} = \frac{22}{15}$

Multiply a fraction by another fraction
Ex.
$\frac{2}{3} \times \frac{5}{5} = \frac{2 \times 5}{3 \times 5} = \frac{10}{15}$

Add two fractions with unlike denominators
Ex. $\frac{2}{3} + \frac{4}{5} = \frac{10}{15} + \frac{12}{15} = \frac{10+12}{15} = \frac{22}{15}$

Change two unlike fractions to equivalent fractions having a common denominator
Ex. $\frac{2}{3} = \frac{2}{3} \times \frac{5}{5} = \frac{10}{15}$
$\frac{4}{5} = \frac{4}{5} \times \frac{3}{3} = \frac{12}{15}$

Determine the most useful name for *one* to rename a fraction to higher terms, given the fraction and a desired denominator.
Ex. $\frac{2}{3} \times \square = \frac{\triangle}{15}$
$\square = \frac{5}{5}$

Construct different names for one of the form $\frac{a}{a}$
Ex. $1 = \frac{2}{2} = \frac{3}{3} \cdots$

Determine the missing factor, given a product and one factor.
Ex. $3 \times \square = 15$
$\square = 5$

Determine the L.C.M. of two numbers.
Ex. 3, 5

Number	Multiples
3	3,6,9,12,⑮,18 · · · ·
5	5,10,⑮,20· · · · · · · ·

L.C.M. is 15.

Identify elements which are common to two sets. (Intersection set)
Ex. (2, 4, 5, 7)
(3, 4, 5, 6)

Common elements are 4, 5

Name the multiples of a given number.
Ex. For 3:
(3, 6, 9, . . .)

239

Sample Problems for Cooperative Groups

The following tasks or problems are examples of problems which can be given to two or more children to solve cooperatively. Each task focuses on some aspect of a computational procedure. The extent to which each task is truly a problem for children to solve depends upon the knowledge and skills of the particular group. of children.

A. Whole Number Numeration

Show 352 with a set of wooden base ten blocks. Also show the same amount of wood using one less ten rod. Explain how you know the two sets of blocks contain the same amount of wood.

(Similar tasks can be designed with decimals.)

B. Whole Number Multiplication and Division

1. Find the missing digits in each of these.

$$
\begin{array}{r}
8\ 7 \\
\times\ 7\ 4 \\
\hline
\square\ \square\ 8 \\
\square\ \square\ 9 \\
\hline
6\ 4\ \square\ \square
\end{array}
$$

$$
6\overline{)\begin{array}{l}7 \\ \square\square \\ \square\square \\ \hline 0\end{array}}
$$

$$
7\overline{)\begin{array}{l}\square 6 \\ 1\ \square 2 \\ 1\ 4 \\ \hline \square\ 2 \\ \square\ 2 \\ \hline 0\end{array}}
$$

2. Solve this puzzle by finding missing digits. **Solve the puzzle in at least two** different ways.

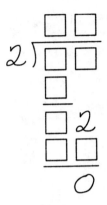

C. Fractions

1. Find the missing number.

$$\frac{3}{8} + \frac{\square}{8} = \frac{1}{2}$$

2. In two different ways, show that

$4\frac{1}{4} \times 4$ does **not** equal $16\frac{1}{4}$